The Socialist People's Car

The Socialist People's Car: Automobiles, Shortages, and Consent on the Czechoslovak Road to Mass Production (1918-64)

Valentina Fava

Amsterdam 2013

This publication is made possible by Eindhoven University of Technology and the Foundation for the History of Technology.

ISBN 978 90 8964 399 5
e-ISBN 978 90 4851 563 9
e-ISBN 978 90 4851 564 6 (e-pub)

Design and typesetting: Ellen Bouma, Alkmaar, the Netherlands

Cover image: Octavia, rodinna 1, piknik, 985 (produced between 1959-64) or Felicia obr. 52

Amsterdam University Press, Herengracht 221, NL-1016 BG Amsterdam
www.aup.nl

Table of Contents

List of Tables

List of Illustrations

List of Abbreviations

AEC	American Engineering Council
AIC	Automotive Industrial Council
ASAP	Akciová společnost pro automobilový průmysl (Joint Stock Company for the Automobile Industry)
AVIA	Letecké závody, národní podnik (Aviation Works, national enterprise)
AV ČR	Archiv Akademie věd České republiky(Academy of Sciences of the Czech Republic)
AZNP	Automobilovè závody, národní podnik (Automobile Works, national enterprise)
CIOS	Comité International de l'Organization Scientifique du Travail
COCOM	Coordinating Committee for Multilateral Export Controls
COMECON	Council for Mutual Economic Assistance
ČAZ	Československé automobilové závody (Czechoslovak Automobile Works)
ČNSP	Česká strana národně sociální (Czech National Socialist Party)
ČSDP	Československá sociální demokracie (Czechoslovak Social Democratic Party)
ČSKO	Československý národní komitét pro vědeckou organisaci (Czechoslovak National Committee for the Scientific Management)
ČSL	Československá strana lidová (People's Party)
ČVÚT	České vysoké učení technické (Czech Technical University)
ČZAL	Československé závody automobilové a letecké, národní podnik (Czechoslovak Automotive and Aviation Works, national enterprise)
ČZLK	Československé závody lehkého kovoprůmyslu, národní podnik (Czechoslovak Works for Light Engineering, national enterprise)
ČZKS	Československé závody kovodělné a strojírenské, národní podnik (Czechoslovak Works for Engineering and Steel Production, national enterprise)
ČZTS	Československé závody těžkého strojírenství, národní podnik (Czechoslovak Works for Heavy Engineering, national enterprise)

ČZVK	Československé závody všeobecné kovo výroby, národní podnik (Czechoslovak Works for General Engineering and Steel Production, national enterprise)
ČZVV	Československé závody na výrobu vozidel, národní podnik (Czechoslovak Works for Motor Vehicle Production, national enterprise)
DS	Demokratická strana (Slovak Democratic Party)
FEA	Foreign Economic Administration
HR ÚV KSC	Hospodářská rada, ústřední výbor, komunistická strana Československa (Economic Council, Central Committee of the Czechoslovak Communist Party)
KOMSOMOL	Kommunisticeskij Sojuz Molodezi (Soviet Youth Organization)
KSČ	Komunistická strana Československa (Communist Party of Czechoslovakia)
KSS	Komunistická strana Slovenska (Communist Party of Slovakia)
LIAZ	Liberecké automobilové závody (Liberec Automobile Works)
Map	Ministerstvo automobilového průmyslu a zemědělských strojů (Ministry of Automobile Industry and Agricultural Machinery)
MAP	Masaryková akademie práce (Masaryk Academy of Work)
MS	Ministertstvo strojírenství (Ministry of Engineering)
MVS	Ministerstvo všeobecného strojírenství (Ministry of General Engineering)
NAMI	Научно Автомоторныи Институт (Scientific Research Automobile and Automotive Engines Institutes)
NEP	Новая Економическауа Политика (New Economic Policy)
NOT	Научнауа Организциа Труда (Scientific Organization of Work)
NSDAP	Nationalsozialistische Deutsche Arbeiterpartei (National Socialist German Workers Party)
OEM	Office of Emergency Management
OPM	Office of Production Management
OTK	Отдел техничесово контрол (Department of Technical Control)
PAL	Spojené závody pomocného automobilového a leteckého průmyslu, národní podnik (Associated Factories for Motor Vehicles and Aviation Parts)
PIMCO	First International Scientific Management Conference
REFA	Reichsausschuss für Arbeitsstudien
ROH	Revoluční odborové hnutí (Revolutionary Union Movement)

STU	Slovenská technická univerzita v Bratislave (Slovak Technical University in Bratislava)
SÚA	Státní ústřední archiv (National Central Archives)
SÚP	Státní úřad plánovaci(State Planning Office)
SVÚM	Státní výzkumný ústav materiálu(State Research Institute of Materials)
UNRRA	United Nations Rehabilitations and Reconstruction Administration
ÚPV	Úřad předsednictva vlády (Secretariat of the President of the Government)
URO	Ústřední rada odborová (Central Union Council)
ÚVMV	Ústav pro výzkum motorových vozidel (Motor Vehicles Research Institute)
VHJ	Výrobní hospodářské jednotky (Production Economic Units)

Introduction

This book aims to contribute to the historiographical debate on the circulation of technology and organizational knowledge in the automobile industry in the automobile industry in twentieth-century Europe. In particular, it examines the transfer of the Fordist production model from the United States to East Central Europe, using Škoda Auto, the leading Czechoslovak car manufacturer, as a case study. In 1991, Škoda was acquired by Volkswagen and today is one of the few companies of socialist Europe to have retained, within its new group, not only its own brand identity but also its headquarters and production facilities.

The focus of the research covers the fifty years between 1918 and 1968, a period marked by both the growing influence of the American model of mass production on the European automobile industry and the transformation of the economies and societies of Central and Eastern Europe in line with the Soviet model of a planned economy and industrial organization. The efficiency and innovative capacity of enterprises formed the battleground for the clash between two opposing socioeconomic systems and two different ways of interpreting mass production.

While the initial industrialization of the Czech lands was influenced by the German industrial model, by 1918, Czech managers and engineers involved in the automotive industry, like those in the rest of Europe, were increasingly attracted to what was generally seen as the new "gospel of productivity": the American model of mass production. However, after 1949, the Czechoslovak industry entered a phase of modernization based on management principles and technical practices drawn from the experience of Stalin's Soviet Union. From the 1950s onwards, these two different approaches to mass production were concretely juxtaposed in Czechoslovak factories in the form of production practices. At the same time, in both the technical press and the documents issued by ministerial authorities and planning agencies, they were at the heart of a heated debate over organization and management. However, the application of the Soviet model of economic management to the Czechoslovak industry had disastrous results, which the reforms of the 1960s attempted to rectify.

The Czechoslovak case raises the question of how the stratification of technical and organizational knowledge and technology shaped production practices in the Czechoslovak industry while also helping to form a specific technical and organizational culture. This book will show us how this culture was formed, what its components were, and how it managed to continue to survive after 1948, despite the impositions of a repressive and authoritarian system. The many political and institutional transitions undergone by Czechoslovakia in the recent past are an invitation to examine exactly how the engineers and technicians, who somehow personified this culture, related to the Party's power and to socialist ideology.

In this perspective, the Czechoslovak case offers many opportunities to scholars interested in industrial and technological relations between the two Europes during the Cold War period. It also reflects on the contribution that an analysis of business history (including a cultural dimension) can make to the debate on transition.

In order to provide answers to these questions, this work reconstructs the events that affected Czechoslovak automobile production, underscoring the organizational and technological changes affecting Škoda's facilities and the ideology of the technicians working in the automotive field at various levels of the socialist economic bureaucracy.

In order to fully understand the transition and formation of a specific technical and organizational culture in Czechoslovakia, we will look at three crucial phases:

1. The initial contact with and re-elaboration of the "American model of mass production", which occurred mainly during the First Republic (1918-38) and in the first few years after World War II (1945-48), leading to the concept of a "people's car and a people's technology for a people's democracy";
2. The introduction of the "Soviet system of economic management" in Czechoslovak factories, following the launch of the First Five-Year Plan (1949-53), and the attempt to create a specific "socialist technology" and "socialist artifact" ideologically and concretely (1953-63);
3. The period of reforms (1963-68), when the attempt to "hybridize" and adapt the Soviet model to local production conditions shifted from a technical debate on industrial efficiency at the factory or ministerial level to assume overt political overtones. Although the reform movement was defeated after 1968, the automobile returned as the main "consumer good" and once again became a priority for planners and an important instrument in the Czechoslovak Communist Party's new search for legitimacy.

The main arguments of this study follow four chief lines of research: the concept of adopting and re-elaborating or hybridizing foreign production models by local actors has not been examined in great detail for Central and Eastern Europe. Studies on the "Americanization" of European industry have largely concentrated on the countries participating in the Marshall Plan, while the prevalent view of the process of "Sovietization" in Central and Eastern Europe has underscored the imperial nature of transferring the Soviet system of industrial organization, overlooking the possibility of a critical re-elaboration by local actors.

This study opens up new perspectives on the above concept, showing how from the 1920s onwards, Czechoslovak technical experts were particularly active in re-elaborating inputs from their American counterparts in an attempt to adapt them to the Czechoslovak production and market scenarios. It also shows how, after 1948, Czechoslovak technical experts applied this same process of selection, reinterpretation and adaptation to the so-called Soviet system, trying to maintain commercial and technical contacts with overseas industry and integrate the input from the capitalist industry (technology and organization) into the Soviet industrial projects.

The ideas of socialist organization and technology that guided the development of the Czechoslovak automobile industry were thus the product of the dictates of Soviet organization and the limitations imposed by the shortage economy on the one hand, and of a continual, uninterrupted comparison with the American "one best way" and with the latest developments in the worldwide automobile industry on the other. The data on the Czechoslovak automobile industry clearly show that commercial and technological relations across the Iron Curtain during the Cold War were considerably more intense and continuous than traditional historiography would have us believe

This book shows that the expertise and know-how of the technicians were only partly able to influence production or affect the enterprise strategy. Indeed, in socialist enterprises, which for most of their history were shorn of decision-making powers regarding production, distribution and structure, individual technical and organizational "competencies" were prevented from developing into the Chandlerian organizational "capabilities" essential to the performance and survival of a capitalist company. The history of the Czechoslovak automobile industry is therefore characterized by an imbalance between the technological backwardness of the manufacturing process and products and the wealth of technical and organizational knowledge the industry had at its disposal.

The quantitative and qualitative data on Škoda's production, productivity, and product quality and the comparison with the corresponding results in Western automobile companies testify to the very difficult conditions that existed for production and technology throughout the period in question. Neither the investments made in the 1960s, which permitted plant modernization and the launch of a new vehicle, nor the informal practices developed to deal with the bottlenecks resulting from the planning process were sufficient to close the gap in production and technology with the worldwide automobile industry that had emerged after 1949. On the contrary, the investments that marked the end of the "building socialism" phase only highlighted the contradictions between the promises of socialist ideology and the reality of the factories.

The existing legacy of technical and organizational knowledge did not immediately translate into practices or innovation, nor did it produce any "innovative hybrids". The informal practices that "greased the gears of the planning mechanism" and the formation of socio-technical networks did, however, create fertile ground for the transition from Fordism to "lean production", a key factor in Volkswagen's acquisition of Škoda.

The conflict between this rich legacy of technical knowledge and the impossibility of achieving production efficiency because of technological and organizational shortcomings has also been analyzed in terms of its political consequences, especially regarding its effect on the Czechoslovak technical experts' adherence to socialist ideals and on the process of de-Stalinization in Czechoslovakia.

An analysis of the strategies employed by the actors and their long-term relations has led to some hypotheses regarding the motivations behind the Czechoslovak technical experts' critical but insistent and at times risky search for knowledge and the impact that contact with Western efficiency had on the world of Czechoslovak production.

This study shows that starting in 1918, there was already a clear objective of a technocratic and nationalist nature behind the re-elaboration of foreign technical and organizational practices on the part of Czechoslovak technical experts. Although echoing certain aspects of the "Czechoslovak path to socialism", it differed in its productivist nature. From the time of the First Republic and the initial contacts with the "American model of mass production", a strong connection was established between industrial modernization and the construction and survival of the institutions of the Czechoslovak state. This connection was subsequently reinforced by the Treaty of Munich (1938) and the nationalization of strategic industries (1946). In 1948 and then again in 1952-53, the call to build socialism and create socialist technology appealed to these nationalist and technocratic tendencies.

Thus, for the technical experts, ensuring the efficient functioning of factories with access to the latest knowledge and technology – regardless of its origin – represented a way not only to guarantee an enterprise's performance but also to safeguard the very survival of Czechoslovakia as a nation.

However, starting in the early 1960s, the practical limits of "socialist technology in use" became self-evident: despite plant modernization and the increased resources available to industry, the planning shortcomings continued to put the continuity of production flows at risk, while concurrently damaging plant productivity, the professional reputation of Czechoslovak technical experts, and their national pride.

The automobile and its function in socialist Czechoslovakia represent the main thread of this work: our study treats the automobile as "product", but it would be disingenuous to overlook its other attributes. While in Western Europe (and elsewhere) the car symbolized economic and social modernization, in socialist Europe it was, more than anything else, an "object of desire". Isolation from the capitalist West, reduction in private consumption and the priority given to military and strategic production forced the automobile industry of socialist Europe to develop in a different context, one characterized by low production volumes, insignificant allocations of capital, sparse investment and reduced markets. This book examines the fundamental steps, reconstructing the development of the automobile's "official" function: people's car, military vehicle, export item and symbol of the well-being promised by socialism. Finally, this study shows that in a socialist country, the automobile was an inexhaustible source of contradictions: as an item to be exported in exchange for hard currency, the Czechoslovak automobile forced planners and technical experts to come to terms with the enormous productivity of Western automated systems and the increasingly aggressive marketing strategies adopted by their competitors. Given that it was seen as a socialist people's car, the automobile reminded people of the initial wave of motorization in Czechoslovakia under Masaryk and represented a promise that the planners were incapable of keeping.

From a methodological standpoint, this work combines two approaches. Firstly, it reconstructs the development of the company structure, the changes in production technology, and the adoption of methods of work organization borrowed first from the U.S.A. and later from the USSR, with reference to developments in other European and worldwide automobile industries. Secondly, in analyzing the technical and travel reports written by engineers working at various levels of bureaucracy, this study attempts to shed light on the reflections of the actors involved, the models they imitated and goals they attempted to achieve, by applying research tools commonly used in ethnography and microhistory. Along with the questions typical of company cases in business history (relative to company structure, performance and technology, as well as production levels and product competitiveness), it addresses political and social questions concerning the degree to which technical experts in one of the most important factories of Central Europe shared the ideals of economic and industrial modernization underpinning the Soviet civilization project for Czechoslovakia.

This study reviews the events that affected the entire Czechoslovak automotive sector, based on material available in company records and national archives. The material has allowed me to reconstruct with a certain degree of precision not only the development of the sector but also the decision-making processes that helped to seal the fate of the automobile industry and motorization in Czechoslovakia at every level of the hierarchical pyramid. The archival sources have been supplemented by secondary sources including automotive magazines and the leading management periodicals.

Many people have aided me during the course of this project which lasted over a period of ten years. I am especially grateful for their thoughtful comments and support to Franco Amatori, Duccio Bigazzi,Yves Cohen, Karen Johnson Freeze, Ruth Oldenziel, Daniele Pozzi, Ferruccio Ricciardi, Sigfrido Ramirez Perez, Johan Schot, Phil Scranton, Lewis Siegelbaum, Ivan Tchalakov and Jonathan Zeitlin. The network Tensions of Europe as a virtual department has proved a congenial location for my work, the Max Weber Program at the European University Institute and the Helsinki Collegium for Advanced Studies provided material and scientific support without which I could not have completed this book.

I owe a special thanks to Eva Martinková, a translator and a friend without whom this project would have been impossible, Brenda Porster, who took care of the English translation of this volume, and Valerie Kidd who meticulously copy edited the manuscript.

I want to acknowledge the intellectual and professional help of Lukaš Nachmann, head of the Škoda Auto corporate archives and I thank Giorgio Bigatti, director of the Isec Foundation, for granting me the kind permission to translate and rework the Italian version of this volume.

I am also grateful to Jan Korsten and the Foundation for the History of Technology in the Netherlands for taking care of the book production process.

This book is dedicated to the memory of Karen Johnson Freeze.

Chapter 1 Scientific Management and Czechoslovak Industry, 1918-38

A New State and a Body Shop Called *America*

The date that best symbolizes the Czechoslovak automobile industry's first encounter with the American model of mass production[1] is 1928. In that year, the merger between one of the oldest Czechoslovak automobile manufacturers, Laurin and Klement from Mladá Boleslav, and the well-known engineering concern, Škoda of Plzeň, bore its first fruit: a new body shop called *America* producing up to eighty-five cars a day.[2]

The *Zprávy Československého národního komitétu pro vědeckou organisaci* (ČSKO) [Bulletin of the Czechoslovak National Committee for Scientific Management] presented the plant as cutting edge in Czechoslovakia's new criteria for the rational organization of production, regarded in the 1920s as the foundation of American economic prosperity. The factory layout comprised two main buildings: the body shop, called *America*, and a machine shop equipped mainly with American and German machine tools. The body shop took up four floors: carpentry was done on the first floor and upholstering on the second, the third was for sheet metal working, while the final stages, painting and finishing, were done on the top floor. The machine shop, 200 meters long and 130 meters wide, produced the chassis. The manufacturing processes were divided into separate groups: engine, friction, gears, frame, and front and back axles were produced and assembled in separate shops. After the parts were run in and checked, they were brought to the assembly line.

The machine shop was organized on a production-line basis, with roller conveyors carrying parts from one machine to the next. The two new buildings were connected in such a way that the finished bodies were transferred from the fourth floor of the body shop to the machine shop, where they were mounted on the new chassis. Smelted castings and forged parts came from the Plzeň steelworks, as did some of the frame components and machine tools. Transport between the two cities was via a private railway connecting the production facilities, which optimized the speed of loading and unloading.

The article in the *Bulletin* described in great detail the planned, uninterrupted progress of products from freight cars to the arrivals warehouse, from the warehouse to the shops, and from one shop to the next. It stressed the fact that movements were guaranteed and controlled by an accurate scientific system of checks and largely mechanized internal transport, as well as by strict Taylorist principles in the division of work.[3]

According to the Taylorists who wrote the article, even if the tasks performed were simple and relatively light, workers still had an important role in controlling production. In contrast, *Škodovák*, the factory newspaper, described the new plant *America* quite differently: "slave shops" similar to Detroit, as hellish as the ones described by Céline, where conflict between the sexes was born and exasperated. Above all, it was an example of the worst sort of Americanism, whose purpose was more the exploitation of manpower than its rational utilization. Workers were reduced to "human figures" who wandered like shadows among machines that were "inferior, broken-down, and inadequate" but that nonetheless could "swallow, eliminate, and substitute" them.[4] Far from serving as a symbol of progress and modernity, *America* was an example of second-rate rationalization. It had borrowed only the worst from Americanization, instead of becoming a source of well-being for workers and owners alike.

The contrasting views of the factory depicted by these different descriptions created unease in the new state of Czechs and Slovaks. Attempts to overcome what seemed to present a threat of social fracture were even greater in Central Europe and in recently formed countries than elsewhere and would characterize Czechoslovakia's industrial and political history. As photographed on the eve of World War II, *America* undoubtedly represented the culmination of processes begun roughly ten years before with the formation of the new state. In its concrete reality, the factory embodied many of the contradictions in the "golden years" of the Czechoslovak Republic.[5]

The practical problems that at the dawn of the twentieth century had made it desirable and necessary for the U.K., France and the U.S. to formulate precise and generally applicable theories of organization, emerged in Bohemia with the birth of the Czechoslovak state.[6] In 1918 an attempt was made to give the "multilingual and heterogeneous territory" that had made up the Habsburg Empire as "rational and homogeneous" a state system as possible. This difficult task guided the complex process of creating political and economic structures that would guarantee the survival of the diverse state entities created by the Treaty of Versailles.[7]

The Habsburg economic order had been founded on an unequal development of the various parts of the Empire: an industrialized west in the Czech lands, which constituted one of the "peaks" of European development, and an agricultural east that fitted into the developmental dynamics of Europe's outlying areas. The passing of this order left the successor states in the various parts of the ex-Empire with the task of correcting this inherited imbalance, making it necessary for each of them to radically restructure its economy.[8]

For Czechoslovakia, the need to take into account the Austro-Hungarian structure of its economy was especially pressing.[9] The new state had inherited the most important industrial plants with 70 percent of its industrial capacity and 44 percent of the total manpower in only 21 percent of the territory. But at the same time it had lost its main markets, located in the less developed areas of the Dual Monarchy. The ability to compete internationally suddenly become vitally important, since more than 30 percent of the newly founded state's industrial production had to be exported. Some entrepreneurs and technicians came to see scientific management as the ideal instrument for the rationalization that would allow them to enter the world market[10] – even more so because the war had led to a noteworthy expansion of the possibilities of mass production. Moreover, the employment of women and common laborers in factories had created an urgent need to raise the capacity of management and coordination to previously unimaginable levels.[11]

From the very beginning, the introduction of new organizational theories and Taylorist-oriented techniques was strongly conditioned by situations and problems related to the origins and structures of the new state. On the eve of the war, travel reports from neighboring Germany already described new forms of organization of production that had started to change the appearance of factories and of work. The debate on Taylorism had already begun. In 1911 a book entitled *Modern Organization of Work in Mechanical Industry* [*Moderní organisace v strojírnách: studie technicko-hospodářská*] was published in Prague. The author, Albín Bašus, was a young engineer who had worked in Germany for many years. There he saw the already vast literature on scientific management and the German translations of the main works of Frederick W. Taylor.[12] In the years immediately before the outbreak of the war, the Czech debate on scientific management grew more intense. The American experience was the central focus of Stanislav Špaček, the first to translate the works of Frank Gilbreth and Hugo Münsterberg into Czech and to spread Taylor's principles through his many writings.[13] Yet, despite Špaček's efforts, the knowledge of Taylor's theories remained limited to the technical and intellectual elite.

It was the military mobilization and reconversion undertaken in the context of the new state system that was to stimulate a systematic discussion on practical applications. An important role in this discussion was played by the Masaryková akademie práce [Masaryk Academy of Work] MAP, an institution founded by Špaček and a group of engineers to promote the principles and practices of scientific management in Czechoslovakia, while championing an interesting reformulation. Alongside the essential activity of "practitioners" who wanted to introduce elements of industrial rationalization and scientific management into factories, MAP's activities also had a political and ideological valence, envisioning a new role for Czechoslovak engineers and technicians in the public sphere. In effect, it constituted a sophisticated attempt to merge Taylor's mental revolution with the ideals of humanitarian socialism of Tomáš Garrigue Masaryk, under whose aegis the Czechoslovak Republic was born.

The project to found an independent institute designed to spread scientific management in the country was presented to Parliament in December 1918.[14] However, the final approval and start-up of activities in January 1920 were the result of Masaryk's intervention and of a donation to the new republic by Czechoslovak emigrants to the U.S., which the President decided to use to finance the newborn institute.[15] Thus, the Academy was born not only out of enthusiasm for the high productivity that the introduction of scientific management seemed to assure but also in a climate of heartfelt admiration, shared by the President, for the model of social efficiency and political freedom that American democracy seemed to embody.[16]

The first phase of MAP's activity (1920-26) took place in close collaboration with the American Engineering Council (AEC).[17] This organization gave economic and intellectual support and assured that MAP had a significant role in the process of socialization and cultural promotion of scientific management taking place in Europe in those years. Indeed, the association of American engineers chose Prague as the seat of the First International Scientific Management Conference (PIMCO) in 1924, and then in 1926 as headquarters of the Permanent Delegation for the Organization of Scientific Management Conferences and seat of the permanent secretariat of the *Comité International de l'Organization Scientifique* (CIOS).[18]

Masaryk Academy and in particular the *Ústav pro technické hospodářství průmyslové* [Institute for Industrial Economics], founded in 1922 by Václav Verunáč, incentivized "industrial pilgrimages" and the development of young Czechoslovak engineers through its untiring organizational and networking activity.[19] At the same time, MAP was also active in introducing the business elites to ideas and practices developed in America.[20] To that end, MAP distributed a questionnaire on the frequency of industrial accidents to the major factories in the country in 1926 and promoted the most important Taylorist campaigns, including one favoring the standardization of tools and public administration reform. The most significant survey, 'Waste in Industry', on losses in industrial production was boycotted by local entrepreneurs, who thought it interfered too much with their managerial practices.[21]

In point of fact, there was a considerable gap between the high theoretical level of the technicians and the real diffusion of scientific management techniques in Czechoslovak companies and factories. The spread of the new organizational methods was hindered by an overly fragmented industrial structure and by the government's tendency to promote protectionist and monetary measures to solve the national industry's problems of competitiveness. The Academy's activities came up against the indifference, if not the outright hostility, of the majority of Czech entrepreneurs, who did not agree with contents they considered too "American", theoretical, and ill-suited to the conditions of local companies, especially small and medium ones.[22]

The Masaryk Academy did not limit itself to sponsoring the unadulterated imitation of the American model. Careful analysis of the activities of various institutes reveals serious reflection on the problems specific to Czechoslovakia and on the ways the American experience could be adapted to suit the complex local situation.

It is significant that MAP continued to focus on issues that valued the human factor and collaboration between workers and employers. They were issues particularly relevant to the specific conditions in this new central European state: in Czechoslovakia, the theories of the "human factor" and psychotechnics mitigated the Taylorist conception of the worker as "plebalia", finding an immediate echo in the key role played by the working classes in the struggle for national independence.

Indeed, MAP's primary goal was to demonstrate the advantages of American techniques within the confines of factory and business management while performing a politically relevant action by showing how the new methods could improve workers' lives by raising both their spiritual level and their material well-being.[23] This is proved by the popularity of psychology and physiology of work studies in Czechoslovakia as far back as 1919. The *Psychotechnický ústav* [Psychotechnical Institute], founded in 1920, was one of the cornerstones of MAP. This explains the lasting bond between the Institute and Frank Gilbreth and his wife Lillian, whose chief work "Motion Study" was translated by the Masaryk Academy in 1923.[24] Similarly, Harrington Emerson, who had an entire issue of *Nová Práce. List pro organizaci práce, techniku a sociální politiku* [New Labor. Journal for the Labor Organization, Technique and Social Policies] published by the *Jednota přátel Masarykovy akademie práce* [Club of the Friends of the Masaryk Academy of Work] dedicated to his work in 1932, carried on a lengthy correspondence with the founders of MAP, which culminated in the translation of his "Twelve Principles of Efficiency" in 1936.[25]

Further confirming the interest of MAP members in theories of the "human factor" and the "humanization" of Taylorism are their contacts with certain Italian exponents of what has been defined as *costituzionalismo di fabbrica* [factory constitutionalism].[26] MAP clearly appreciated the work of Francesco Mauro, promoter of the First International Conference on Scientific Management in 1924 and President of the International Committee of Scientific Management from 1926.[27] Interestingly enough, MAP also appreciated the theories of the Fiat engineers Mario Fassio and Mario Fossati, who wanted to use the principles of scientific management to make the most of their workforce. They were convinced that maximum efficiency could be obtained only through consensus and that scientific management should be implemented through a regime of high wages.[28] This would allow the Italian industry to increase both its competiveness on the international market and the living standard of its workers, thereby calming the climate of 'two red years' of social conflict.[29] These two engineers clearly represented a minority position among Italian Taylorists as well.

The echo of the discussions regarding workers' involvement and recognition of the role of factory councils were also clearly perceived. These were topics dear to German reformist socialism. Initially developed in the last quarter of the nineteenth century, this movement gained renewed interest after the war due to the involvement of factory councils in managing the war effort and business leaders' attempts to stem the tide of the revolutionary movement by granting more worker co-determination.[30]

The desirability of collaboration was a highly topical issue in Czechoslovakia as well. It came in response to the strong social discontent prevalent among workers in the early 1920s and to the fear that the workers' protest movement would lead to revolution. The years between 1918 and 1921 were particularly intense in Czechoslovakia as regards industrial relations and the history of the socialist movement, which was characterized by its marked nationalism and the strong support it gave to the struggle for independence.[31] In the last months of 1918 and for the whole of 1919, the war and widespread poverty meant the Czechoslovak people suffered a lack of bare necessities. Immediately after the war, agricultural and industrial production was less than half of pre-war levels, and high inflation had notably reduced the purchasing power of wages.[32] Nationalistic euphoria had allowed the October Revolution to echo in the national independence struggle.

But once this had subsided, domestic tension became more radical. Strikes – trade union membership continued to rise, amounting to almost two million members (1,979,700 in 1921) – and the risk of civil war and uprisings peaked in 1920. And the conflict between the two historical currents of the Bohemian socialist movement, latent in the years immediately following the war, resurfaced: the reformist current was opposed to the internationalist Marxist current, whose leader, Bohumír Šmeral, went on to found the Czechoslovak Communist Party in 1921.[33]

These were the years that saw the introduction of agrarian reform and the eight-hour workday, political-economic measures with a strong social impact. When workers' discontent awakened fears of a possible revolutionary outcome, MAP came forward as a candidate for the important role of political mediation, offering scientific management as a precious instrument for healing the ongoing social conflict.

The two periodicals dealing with the topic of scientific management, both connected to MAP's activities, clearly illustrate the complementary souls of the Czechoslovak rationalization movement. *Nová Práce* fostered political mediation and propaganda aimed at working-class consent. In contrast, the *ČSKO Bulletin* was decidedly oriented towards the concrete diffusion of American techniques.[34] These two were, however, interdependent, as the composition of the editorial boards of the two publications shows. They were part of a wide-ranging project focusing on linking workers' living standards to Taylorist techniques and national survival. In this context, scientific management seemed to guarantee the material basis necessary for democracy, understood by Masaryk as "the organization of a society founded on work" in which one class could not be exploited by another.[35]

The contents of *Nová Práce* reveal that MAP engineers held very similar views on the organization of democracy. For them, since the factory was the place that fostered collaboration between technicians and workers, it became the key to national solidarity, the material and moral starting point for a healthy, independent, and democratic economy and society. Work had become a moral obligation and, using a military metaphor not unknown to Taylorist propaganda, the factory was the place best suited to carry on the mobilization of the workforce that had begun with the struggle for independence. The discipline invoked was legitimized by a collective goal – increasing the wealth of the nation and safeguarding its independence – and by an infallible scientific method that was *super partes*. Respect for workers' knowledge of their craft on the one hand and collaboration with management on the other seemed the only way to give fair value to workers as possessors of specific tacit skills at the very moment when Taylorism and job simplification put their professional identity at risk. The factory also became a "cell" for implementing the educational program that was one of the goals of Masaryk's democracy, a program deeply rooted in the Bohemian political and philosophical tradition and in the spiritual heritage of Comenius.[36]

In such a cultural framework, there was hardly any space for special or class interests. However, this in no way meant that the social problems of industry or the demands of the working class were to be ignored. Masaryk's theory, which was faithfully followed by technicians, was that the social question was a sensitive issue, and one that had to be dealt with from a humanitarian, not necessarily class-oriented perspective.

The arguments used in many articles in *Nová Práce* virtually paraphrased Lenin in his "Immediate Tasks of the Soviet Government",[37] which stressed the importance of discipline, education, and the material and cultural elevation of the worker. At the same time, however, they declared that socialist technicians had the responsibility to offer a valid alternative to revolution. Even if Masaryk and the technicians had some sympathy for Marxism, their condemnation of what had taken place in Russia was absolute. According to Masaryk, Bolshevism was the

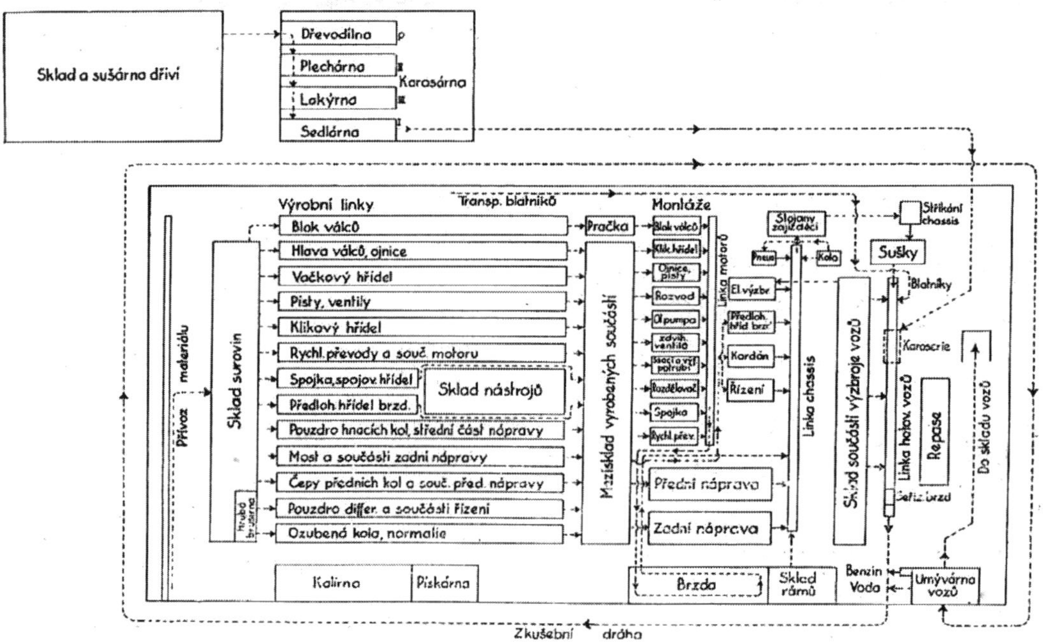

Diagram postupu výroby.

Figure 1.1 – "America", lay out 1931 (Výroba osobních a dopravních automobilů Škoda v Mladé Boleslavi in "Zprávy Československého národního komitétu pro vědeckou organisaci", Prague, February, 1931, n. 2, pp. 17-18)

product of the political culture of Russia, a culture alien to Europe. In his view, Russia's backwardness had led the Bolsheviks to reproduce some of the authoritarian features of the old regime. Conversely, European socialism should lead to a broadening of democracy, not to its denial. The MAP technicians went on to say that raising workers' living standards would not come about through revolution, which would only lead to terror and violence, but through the progressive and socially useful introduction of scientific management, without sacrificing workers' humanity to the needs of production.

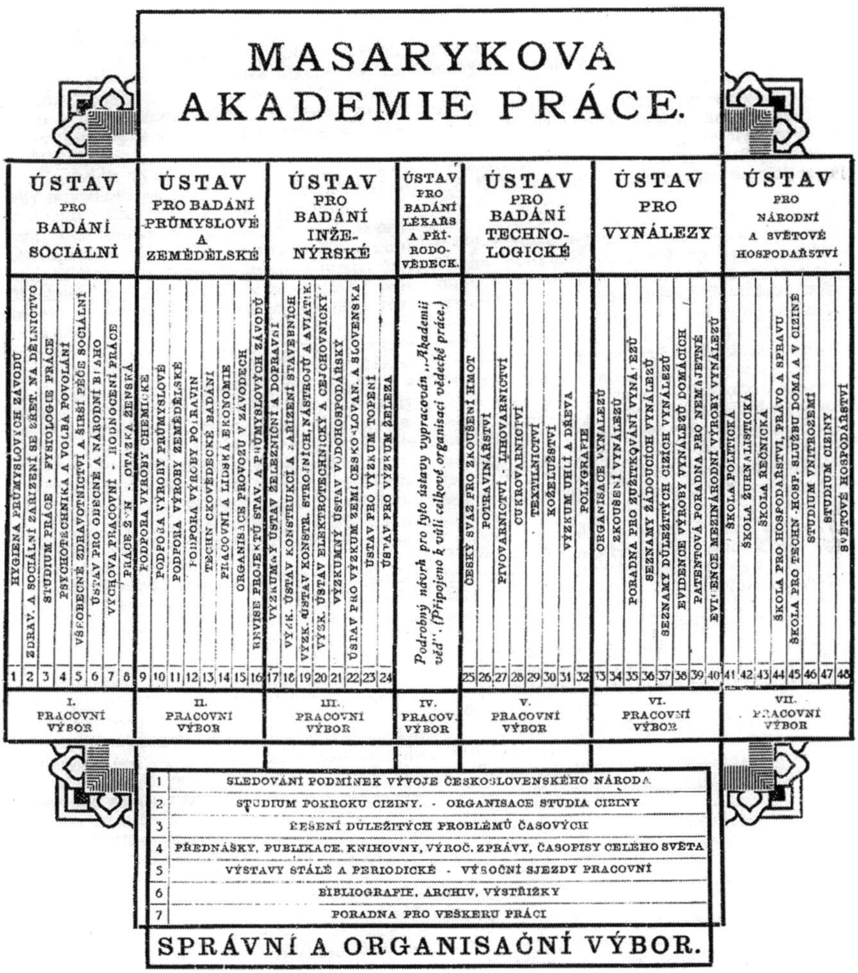

Figure 1.2 – Masaryk Academy of Work, chart

European Roots of Czechoslovak Managerial Americanism

The themes of work and its humanity dominated the MAP engineers' reports of their journeys to America before the great crash of 1929 struck Europe. When MAP translated Henry Ford's biography, *My Life and My Work*, in 1925, more than ten years had passed since Franz Kafka had described the amazement felt by Karl Rossman, a young German from Prague, at seeing the new continent.[38] This wonder for a world invested with features of the imagination became a concrete reality in the MAP engineers' scientific investigations and thirst for learning. In the streets and factories of America they sought out the "secrets" which could serve as the basis for the future prosperity of the new Czechoslovak state.

MAP used its close ties with American engineering associations to sponsor Czechoslovak engineers' travel to America, where they visited the major modern companies.[39] Its work in promoting training continued until the early 1930s when, following the world economic crisis, the Academy found it increasingly difficult to get answers to its requests from America.[40] However, in 1927-28 it was able to collect the reports of the young engineers who had visited America and publish them in two volumes entitled: *Life and Work at Ford* and *Ford and Us.*[41]

It seemed that Fordist philosophy could indeed provide some of the answers the MAP technicians were looking for, to supplement those found in the techniques of scientific management. What seemed to stir their interest and be most readily welcomed was the idea of the system spreading outwards from the Ford factory, which contributed to the creation of a new workforce that shared the company's goals and became its consumers. In fact, more than the mere "move the metal" aspects or the mechanization of production processes, what these young engineers were invited to describe were the industrial relations inside the factory and the extra-economic activities that the Ford factory performed.[42] Thus, even if the assembly line remained the characteristic feature of the Fordist revolution, the one that set the pace of work, many of the specific questions put to the young engineers focused on the quality of work. Those who had worked on the line were asked to describe their sensations and experiences, with the implicit invitation, or so it seemed, to discredit once and for all the political myth that described the Ford workers as "trained apes".[43]

The issue of the simplicity of tasks that could even be done blindfolded, leading to the risk of monotony, the "suppression of workers' humanity", and their consequent alienation, was the thorn in the side of the MAP technicians. They clearly wanted a dialogue with the workers to find a way out of the quandary of the distance created by Taylorism between manual work and the "human content" of labor, as formulated by Gramsci.[44] Yet, the ten engineers' responses were often ambiguous: the heaviness of assembly line work depended on the size of the pieces and the speed of the machines, but it was no greater than traditional modes of work; while the melancholy they felt at times might have been due to homesickness or loneliness, or because they were performing jobs meant for common laborers from the American countryside, not for brilliant engineers from Czechoslovak technical universities.[45]

Far clearer and more reassuring was their evaluation of the climate inside the factory and in the immediate vicinity. The young technicians were struck by the fact that no-one complained about the work or about the system. But what seemed even more surprising was the workers' attitude towards their foremen and vice versa: the friendship, the collaboration, the absence of dispute between the two, whose relations were rife with conflict in the traditional factory, sparked the engineers' curiosity and admiration. The social peace that reigned in Ford shops was possible thanks to the fact that the foremen came from the same world as the workers, while the engineers were familiar with operations, and methods had been determined a priori through diagrams and calculations, following a rhythm preestablished by machines.[46]

Harmony inside the factory was also assured by an efficient safety system, with medical clinics where workers went even for small accidents, as well as by factory outlets where any article could be bought at a reduced price. Internal mobility was guaranteed by professional schools that offered workers the skills needed for promotion. The magazine *Ford News* was another clever instrument to win the workers' admiration for their "inspired" boss, but also for the system in general.

From this point of view, Ford seemed to represent an ideal enterprise for the MAP engineers: social harmony, well-being and efficient production. After they had put in a day's work, workers went back home to a nice, small house paid for in monthly installments, proud of having done their duty. They felt they had contributed not only to their own well-being and that of their families, but also to the well-being of thousands of other workers in America and around the world who worked in factories created thanks to the surplus of their daily work.

Even the MAP technicians risked losing their sense of the fine line between respect for the rules and institutions of the democracy they wished to defend and the insidious paternalistic and corporative scenarios already being institutionalized elsewhere in Europe and which, ten years later, were to find a concrete example not in Ford's America, but nearer to home, in Volkswagen's Germany.[47] Indeed, certain observations, though in the minority, contain disturbing statements of a strictly political nature which, along with the esteem shown for the employer, praised the

American workers' lack of political and trade union affiliations. In his report, a young engineer stressed how the typical American worker spent his leisure time with his family – taking care of his wife and children, the house, and garden – or in church – a cultural and social center. He showed no interest in politics at all; at the poll he voted like his employer, not surprisingly since he was persuaded that the profits of his employers and the fortune of the business were at the heart of his own well-being, granting him his job and high wages.[48]

The Gramscian concept of *coscienza del fine* [goal consciousness] clearly emerged as the hope for the creation of a new type of worker, less difficult than Czech workers.[49] The introduction of a people's car similar to the Model T or to General Motors' Chevrolet was desirable partly because of the incentive a strong automobile industry would provide for related and component parts industries and partly to boost national prestige; but it would also have a positive influence on the "overly politicized character of the Czechoslovaks":

> If they owned a car, they would use their free time taking trips into the countryside and repairing it in the garage rather than on politics and all-too-often useless meetings [...]. The character of our people would also be nobler, and the frequent trips would make them forget life's hardships and sorrows.[50]

Yet, though the economic crisis following the great crash of 1929 made the ideological features and hints of technocracy implicit in the MAP project more and more pronounced, the Academy's activity seems to have remained faithful to the humanitarian ideals and the political project that had originally inspired it until 1938. In Czechoslovakia, where at least since the end of the war the debate on rationalization had been decidedly influenced by the American example, and where decisive changes in the industrial structure were taking place, the great depression was an opportunity to seriously rethink scientific management as well rediscover European management theories.[51] Far from questioning the rational principles of scientific management, the Academy's technicians invested them with even greater hope. Faced with the chaos generated by the economic depression, they reacted with growing ideological rigidity, repeating their faith in the scientific rationality of the method and the need to introduce it in its entirety. The Czechoslovak economy was suffering from the delay in carrying out industrial rationalization. The solution could only come from speeding up the process of rationalization in the context of a planned, controlled economy with public intervention.[52]

In the very years when the economic crisis was having its most drastic effect, MAP published the *Encyklopedie Výkonnosti* [Encyclopedia of Efficiency], a work in three volumes [*Člověk* – People, *Výroba* – Production, and *Obchod* – Trade], which reviewed the history of the theories of scientific management and of the personalities who had created it. This work contained all the key elements of the Academy's previous thinking: faith in this science as a neutral instrument for resolving conflicts of any nature whatsoever, and the importance of cooperation among the parties involved for achieving the greatest possible efficiency. But the *Encyklopedie* placed these elements in a more organic and "encyclopedic" context. The human factor remained at the core of this complex vision, integrated at last within a comprehensive approach to socio-economic development.

In line with the ongoing structural changes, the 1930s witnessed a transition from the exclusion of politics from the world of production to a conscious attempt to apply the principles of scientific management to fields previously outside of the technicians' sphere of interest. In a vision of the economy permeated with ideas of planning, control, and science, technicians took on new responsibilities. Reference to the American technocratic movement and to Howard Scott seemed inevitable, but American technocracy was seen as "ethically colorless" and for this reason not suited to European conditions.[53] Enhancing the ethical dimension of work could counteract the criticisms of "inhumanity" and "venality" directed at Taylorism, which since the 1920s had caused resistance to "Americanization" and rationalization.

This new approach was set out in the so-called Laborethics theory first formulated in 1926, but only systematically expressed in the *Encyklopedie*. *Laborethismus* (Laborethics) was not based on original ideas but was actually a reformulation of ideas that had already largely appeared in the European debate on work organization. According to the Czechoslovak labor ethicists, only ethical techniques and principles of scientific management (operations, analysis, plan, initiative, system, and control), with the collaboration of all economic parties, could guarantee natural progress and development, which of course could not be obtained through revolution. Thus, it was the technicians' duty to spread the concept of ethical work using all the cultural elements at their disposal and to do everything possible to raise people's standard of living and culture.[54] What about the managers? According to labor ethics, they had to understand the pros and cons of work and be capable of involving everyone, right down to the lowest employee (rule of cooperation); to achieve efficiency, they had to be able to investigate even the tiniest details of production processes (principle of analysis), conducting their work while bearing in mind the laws of ethics and professional competence (also applicable to public administration). They had to be aware that, as the world war had shown, technology could just as easily destroy as create, according to whether or not it was accompanied by ethics. This was the challenge of the newborn Czechoslovak state.[55]

The reference to the French and German experience was explicit, especially to the currents of German social democracy and conservative socialism, to Walter Rathenau and Wichard von Moellendorff. The new collaboration between workers and management would find its natural environment in government-controlled cartels and in the mixed economy of the *Neue Wirtschaft*, which Czechoslovakia seemed oriented towards in those years. [56]

Laborethics also called for the entrepreneur's responsibility to the community – in their view, he was the "true organizer of production", and the imperatives of efficiency and general welfare (the principle of public service) required greater moral control on his part, as did the refusal of immoderate profits and speculation.

However, laborethics echoed both Trotsky's speech at the inauguration of the Moscow Thermo-Technical Institute in 1925 and the Czechoslovak National Socialist Party's program of 1930. Distancing itself from the American premises of MAP's activity, it rediscovered *Planwirtschaft* as reformulated by German social democracy in its attempt to institutionalize it, interpreting it as a possible intermediary phase between capitalism and state socialism. Interestingly, we can trace the same matrix in Lenin's writings on scientific management.[57]

But in the second half of the 1930s, the economic equilibrium of Czechoslovakia was changing, while the political climate was degenerating. In 1935 the movement for the autonomy of Sudetenland by Konrad Henlein, who had proclaimed himself the representative of the German minority, won a relative majority of the votes in the elections. At the same time, German capital gradually substituted British and French, making this central European country ever more vulnerable to Nazi ambitions of expansion. [58]

Still, Czechoslovakia remained democratic. The editors of the *Encyclopedia* worked within this context, supporting positions consistent with the politics of Eduard Beneš and the government, which aimed to create a Czechoslovakia that would follow Masaryk's teachings and be "an oasis between fascism and absolutism". Unfortunately, this political position was destined to collide first with international diplomacy and then with Hitler's troops.[59]

Czechoslovakia and Fordism

In Czechoslovakia there were vigorous and fertile debates on scientific management between the two world wars. The Masaryk Academy of Labor quickly

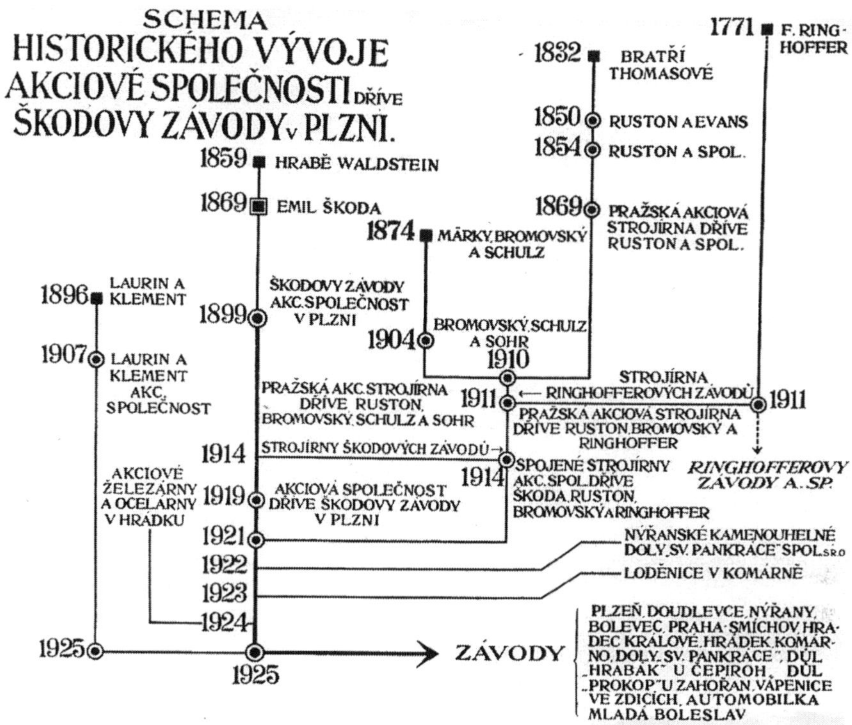

Figure 1.3 – Škoda, chart (Nová Práce, n. 7, *1925)*

assimilated the new ideas from the U.S. and the rest of Europe. Yet, there was a wide gap between the high level of the technicians' theoretical formulations and the actual extent of scientific management methods in Czechoslovak industries. While the MAP featured the first echoes of the Human Relations School, the problems in factories due to lack of information about how to rationalize production and the failure to introduce new machinery and management techniques had not yet been solved.

There were, however, some significant exceptions – particularly enlightened entrepreneurs or managers who were ready for key elements of the American model following the technicians' recommendations. Tomáš Baťa was undoubtedly the most well-known of these. But the Czechoslovak company best positioned to become the model for big business and to exemplify the advantages of the correct introduction of the new management methods was Škoda of Plzeň.[60] An article in *Nová Práce* of 1925 defined it as the "factory of economic and social progress" and traced its history back in time, celebrating its merger with Laurin & Klement, an automobile manufacturer based in Mladá Boleslav, and its arrival in the automobile sector.[61] The body shop, *America*, was in fact the most advanced product of a strategy aimed at faithfully following the developmental logic of the new American model of mass production; the fact that it was a plant designed to produce automobiles increased its symbolic inspirational value.[62]

In the years immediately after the war, the Empire's so-called "monarchy of weapons" risked liquidation, because it lacked not only the financial resources needed for reconversion but also a market that could absorb its products.[63] Škoda found itself in an extremely tight financial situation: the Austrian military administration had not paid its debts, and there were liabilities caused by huge investments in wartime production. *Nová Práce* reported that Škoda had overcome the serious crisis threatening its survival by applying the watchwords from across the ocean: diversification of production in key sectors of the second industrial revolution, vertical and horizontal integration, and attention to workers' standard of living.

There was also a problem regarding ownership. Škoda, originally a foundry belonging to the aristocratic Wallenstein family, was bought in 1869 by Emil Škoda, its chief manager. Educated at the Technical University of Prague and with considerable experience in Germany, he soon turned it into the most important metallurgic and engineering company in the Empire. In 1899 it became a joint-stock company. With the birth of the new state, the company, whose capital was held by the founder's son, Karel, an Austrian citizen, and by German banks, was nostrificated.[64] Its legal headquarters was transferred from Vienna to Prague, and the board of directors was dismissed. After difficult negotiations, which even considered nationalization, the French group *Schneider et Creusot* gained control through a holding company, the *Union Européenne Industrielle et Financière*, created in 1920 to administer banking and industrial holdings in the former Austro-Hungarian Empire.[65]

The new board of directors was elected in May 1922 and included František Hanuš, Josef Havránek[66] as technical director, and Karel Loevenstein[67] as managing director. In line with the most up-to-date trends in international capitalism, it opted for the concentration and diversification of production. They would achieve this by acquiring the majority of parcels of shares in domestic and foreign companies in sectors such as electric and civil engineering (Spojené strojírny, Praha), shipbuilding (Komárno), mining (Sv. Pankrác coal mine), and steel (Hradec Králové). The acquisition of Laurin & Klement was deemed necessary to expand the motor vehicle department, which up to then had mainly produced the luxurious Hispano Suiza on license.[68]

Following its entry into the sector symbolizing mass production, Škoda organized times and methods offices to study the rules of production according to Taylor's principles and to try out systems of incentivized pay. They standardized parts and tools. However, according to *Nová Práce,* the great success of Škoda was not only thanks to competent management, but also the result of the "harmonious collaboration between employer and employee". Škoda was in fact able to reconcile the requirements of company growth with its social commitments, guaranteeing its employees high wages as well as social and housing assistance, and showing it could promote their welfare even in hard times.[69]

Nová Práce's description of the growth of this great metalworking and engineering company almost slavishly imitated reports coming out of Detroit at the time, but with ideological emphasis and jubilation. Ultimately, it ignored not only the intricate network of interests, including political and diplomatic ones, that had made it possible for Škoda to grow into a horizontal oligopoly, but also the difficulties involved in transferring the Fordist model to Czechoslovakia.

Another document that complemented this description, though only reflecting the new automobile company's prospects, was the industrialist Václav Klement's report of his journey to the U.S.[70] In 1895, together with Václav Laurin, Klement had founded Laurin & Klement at Mladá Boleslav as a workshop for assembling bicycles from mainly British component parts. It established itself on the national bicycle market and soon expanded into producing motorcycles. Its transition to automobile production came in 1905, and in 1907 Laurin & Klement became a joint-stock company. It grew rapidly and created sales branches abroad and started producing airplane engines. In 1914, on the eve of World War I, Laurin & Klement had 1,715 employees and a production area of 31,000 square meters. But it, too, faced serious problems of reconversion after the war. At that time it produced thirty models of cars (for only 355 units produced in 1922) and other vehicles, and ten different types of powered engines. From the point of view of production methods, its approach was decidedly one of specialized artisanship. Characterized by high expenses and high selling prices, it was increasingly threatened by foreign and domestic competition. In those same years, the company ČKD, under the *Praga* brand name, was introducing assembly line and management techniques geared to larger series production.[71] Together with the need to enlarge plants through new investments while reducing the range of models, these mounting difficulties convinced the board of directors of L&K to accept Škoda's buy-out.

In 1927, once the agreement was reached, Václav Klement travelled to the U.S. from July to November as a board member of the new company to make personal contact with overseas producers and to negotiate the purchase of machinery and possibly also component parts needed to update the company's automobile production.[72] In his travel report, Klement returned to the topics repeatedly discussed in *Nová Práce* and by MAP. Like Havránek, he had been sympathetic to MAP and had helped to finance it, but his view of the American reality was far more critical and pragmatic. This time, along with an undoubted admiration for American factories, something new was emerging: besides the intellectual debates and expectations of technicians who were experts but also had ideological leanings, what was now felt to be crucial was not so much the search for an industrial or social model as concrete solutions that would allow the Mladá Boleslav plant to work efficiently in its own right.

We can deduce what Klement was really looking for in America from the sample of companies he visited, which were chosen by the Taylor Society. In a memorandum, they recommended to Klement to concentrate on "medium size whose volume of business and management problems correspond closest to those in Europe". That is, companies which were able to survive in the competitive environment of the American automobile industry largely thanks to the excellence of their management. The Taylor Society also asked Klement to take note of the "unusual conditions" found in plants exemplifying mass production based on huge sales volume, such as General Electrics and General Motors: "continuous flow through single purpose machines".[73]

Klement's itinerary re-traced the route of American specialized production. He had three main destinations: Cleveland (for materials and machine tools), Detroit (large and small assemblers – General Motors, Dodge, Packard), and Akron (tires). In terms of sheer numbers, he visited nineteen assemblers, thirty-seven part manufacturers, twenty-six machine tool producers, thirteen body makers, and a number of foundries and related shops (Murray, Lycoming, Buick).[74] After observing the machinery, assembly line mechanization, and the enormous productivity of the single purpose machines, Klement did not fail to notice that all this was the patrimony of only one of the two groups of American automobile manufacturers, that is the group including Ford, Chevrolet, General Motors and Dodge. In contrast, the other group had less "enormous" production and used the new hardware – motorized conveyors and single purpose machines – only in certain phases of manufacture and assembly.[75]

Klement's detailed report stressed various aspects of American factories, often showing more interest in smaller manufacturers whose production and equipment were more like the Mladá Boleslav plant. Specialization seemed to him the "most significant feature of the American automobile sector", and from this point of view the most interesting examples were the producers of component parts and machinery that supplied the great assemblers.[76] Klement emphasized the differences between what he had seen in America and what existed in Czechoslovakia. The evident differences were not just in the final manufacturing processes but also in the quality of the materials and semi-finished products, the components and their supply, the design and development of models and, above all, the quality of the workforce.

However, Klement did not suggest forming a new, more controllable workforce to counter the skilled but rebellious Czechoslovak workers. Rather, he viewed the replacement of manual work that was "expensive and overly dependent on the worker's precision" with machines and the simplification of tasks as an inevitable and ingenious – but not necessarily desirable – solution to a problem specific to American industry: the difficulty of finding capable and skilled workers like those readily available in Mladá Boleslav. Even the welfare measures – housing, canteens, doctors, insurance – that the American factory offered its employees were in reality an attempt to solve the high turnover problems crippling American enterprises. In the same way, distribution systems, forms of advertising, and installment purchasing were seen in the context of the American situation and reconstructed in what we would call an historical way.[77]

There was also a rather curious comparison with the "most Americanized" enterprises in Europe: concerning worker productivity and wage systems, Klement reported a comment by the president of Marmon. In his view, the reason why Fiat had lower productivity and Germany's Opel had higher expenses than American plants, even though both copied American methods and could count on lower salaries, was paradoxically the curbing effect of low wages. But it would seem Mladá Boleslav did not run these risks.[78]

Klement's ideas about what he saw in American factories were strikingly different from the hagiographic tone in *Nová Práce*. This difference may suggest just how wide the gap between the ideal and the reality of Czechoslovak automobile production really was, and it helps us to view *America* from a more accurate perspective.

The construction of the new factory came late compared to the first cycle of reorganization in European auto plants, inaugurated by Fiat's Lingotto as far back as 1916 and inspired by Highland Park. Despite this, *America* did not follow the model of plants built in the years before and after the world economic crisis, which looked to River Rouge.[79] *America* was born when Ford was already designing the Dagenham factory in the U.K. and the Colonia factory in Germany, and while Citroen was creating the *chaine unique*, installed in Javel in 1933. Meanwhile, in the reorganization of its Cowley plant, Morris implemented progressive production, which encompassed total mechanization of the body shop with a network of twelve thousand aerial conveyors.[80]

Though equipped with American machinery and an assembly line and inspired by overseas innovations, the 1928 factory neither had nor aimed to have the dimensions or production capacity of plants that followed the American model more closely. However, this did not prevent Škoda Auto from becoming *Akciová společnost pro automobilový průmysl* (ASAP) [Joint-Stock Company for the Automobile Industry] in 1933. ASAP's headquarters were not located in Plzeň, to emphasize that its requirements in terms of management, resources, and production tools, being aimed at large series if not mass production, were different from what was suitable for the small series or batch production typical of the heavy machine and steel plants in Plzeň. To further stress the break with the craftwork tradition of Laurin & Klement and of the Škoda Works, in 1931 the production of airplane engines was finally transferred to aircraft producer Avia based in Letňany and part of the Škoda concern. At that time, they rationalized the plant's production plan by limiting production to three models of automobiles and two of trucks. Simultaneously, the idea of a "people's car" began to take shape.[81]

Though they viewed mass production with interest and admiration, what Škoda managers needed was a system that could harmonize the Czech tradition of artisanship and highly qualified workers with the problems of obtaining raw materials and existing market conditions. In the end, it was the latter factor that determined the limited share Fordist production would have in Czechoslovakia. In 1928, only 40,000 automobiles were registered, and total annual production hardly reached 13,000 vehicles.[82] Production was divided among ten automobile companies, which produced thirty-three models (eighteen types of cars and fifteen commercial vehicles), so on average no company produced more than 700 automobiles a year per model.[83]

Other obstacles made it difficult for the three main automobile companies (Tatra, Škoda, Praga) to carry out a process of industrial concentration and specialization limiting the number of models they produced. They included a tax penalizing car owners and fuel consumption, as well as strongly protectionist trade measures, which were reinforced in response to the 1929 crash.[84]

The effects of the economic crisis on the Czechoslovak automobile industry were profound. In 1933 Škoda's production fell to less than half of what it had been the year before, and only 36 percent of the staff employed in 1930 were still working at Mladá Boleslav in 1933.[85] But the economic recovery in the second half of the 1930s seemed to open up new prospects: in 1934 Škoda launched the 420 Popular, the first Czechoslovak people's car, while the government started to look more favorably on the spread of motorization in the country.[86] In 1936 Škoda held

Table 1.1 – Motor Vehicle Production (in units) per year in Czechoslovakia, 1898-1945

Year	Automobiles	Trucks	Buses	Total
1898	1			1
1899	7			7
1900	19			19
1901	13			13
1902	12	5		18
1903	13	5		3
1904	3			
1905				
1906				
1907	230			230
1908	260	5	11	276
1909		2	10	452
1910	350		4	354
1911	350	20	5	375
1912	450	10	9	469
1913	600	50	7	657
1914	500	110	2	612
1915	450	550	7	1,007
1916	350	550	10	910
1917	400	1,000		1,400
1918	300	400		700
1919	450	150		600
1920	400	450	1	851
1921	400	300		700
1922	700	300	6	1,006
1923	1,300	350	4	1,654
1924	2,000	450		2,450
1925	2,800	900	1	3,701
1926	5,500	1,800	200	7,500
1927	3,400	1,900	150	5,450
1928	9,100	3,600	400	13,100
1929	10,200	5,200	300	15,700
1930	7,700	4,000	250	11,950
1931	9,000	2,300	200	11,500
1932	7,100	900	100	8,100
1933	5,600	800	20	6,420
1934	8,000	500	20	8,520
1935	9,000	600	30	9,630
1936	11,700	3,300	10	15,010
1937	13,300	2,700	100	16,100
1938	13,900	4,800	100	18,800
1939	10,800	3,900	30	14,730
1940	3,600	4,100	10	7,710
1941	4,200	3,700	2	7,902
1942	4,100	2,400	100	6,600
1943	3,900	2,200	100	6,200
1944	1,700	1,400	150	3,250
1945	300	1,000		1,300

Source: Škoda-Auto Historical Archive (Aša), Mladá Boleslav, Automobilové závody, národní podnik (AZNP), b. 34, Český statistický úřad, zpávy a rozbory, rok 1970, *Rozvoj automobilového průmyslu v ČSR*, tab. 11.

Table 1.2 – Motor Vehicle Production (in units) at Mladá Boleslav Plant, 1907-45

Year	Cars	Trucks	Year	Cars	Trucks
1907	208		1924	478	
1908	239		1925	596	
1909	382		1926	902	
1910	287		1927	1,369	
1911	325		1928	2,000	
1912	460		1929	1,549	
1913	434		1930	3,104	
1914	311		1931	2,536	
1915	255		1932	1,652	
1916	288		1933	432	
1917	122		1934	2,198	
1918	112		1935	2,334	
1919	252		1936	3,631	
1920	147		1937	5,719	1,203
1921	201		1938	6,485	1,109
1922	160		1939	5,624	667
1923	263		1945	2	645

Source: Škoda-Auto Historical Archive (Aša), Mladá Boleslav, Automobilové závody, národní podnik (AZNP), b. 56, *Historický vývoj výrobního profilu závodu*.

the top position on the domestic market with 3,000 cars sold, a figure that was to grow even higher the following year. Mladá Boleslav had six thousand employees and turned out 7,677 vehicles, of which 5,000 were the Popular.[87]

Though Škoda's director, Václav Matouš, was sent overseas to Lycoming in 1937 to purchase new engines and machinery,[88] in the summer of 1938 it was the work organization methods employed by medium-sized French, Belgian, and Swiss enterprises that demonstrated "a system for checking individual output" and "an incentive system that can guarantee the interest of workers and foremen".[89] The last journey Škoda technicians took before the occupation was, in fact, intended to study what was considered a model that succeeded in combining good production organization with both an adequate wage system and planning. Switzerland was thought to be the country most like Czechoslovakia in view of its economic structures. With few exceptions, in plants where craftsmanship and hourly pay prevailed, the technicians noticed intensive organizational and planning activities formerly associated with time and piecework measurement.[90] Thus, after the many journeys to America, it was Switzerland that in 1938 represented the final stop: they seemed to have found the benchmark against which to measure Škoda's prospects for growth and organization in the industrially small-sized confederation which remained neutral but was able to defend itself.

Figure 1.4 – Popular 927, prospekt žena s větrem ve vlasech – Women with wind in her hair (Škoda Auto Muzeum)

Scientific Management and War Economy

After annexing the Sudetenland in September 1938 on the basis of the agreement stipulated in Munich, on 15 March 1939, the German army put an end to the Czechoslovak Republic. While Slovakia declared its independence and instated a collaborationist government, what was left of the state of the Czechs and the Slovaks became part of the Protectorate of Bohemia and Moravia, a new province of the Third Reich. Between 1939 and 1943, the German administration granted formal partial autonomy to the government of the Protectorate, following the model of some colonial experiences. But this autonomy was gradually more and more limited as the war progressed, and was finally totally suppressed in September 1943.

From the ethnic and political standpoint, a policy of "denationalization" and "Germanization" of the Czech population was carried out with the collaboration of the Union of the Germans of the Sudetenland, which had merged with the National Socialist German Workers' Party (NSDAP). For the economy, the occupation meant the subjugation of factories to the needs of German industry and the complete militarization of the economy of the Czech and Moravian territories, in a rising trend that followed the development of military events.[91]

According to Hermann Göring's plans, the zone occupied by the Germans was to become a homogeneous economic area totally coordinated from Berlin, and with reciprocal obligations. To achieve this goal, they used various financial and political instruments.[92] The state now controlled all sectors of production, distribution, and consumption, confiscating the National Bank's gold reserves and Czech foreign credits. The banking system was completely reorganized, and commercial banks were liquidated. The customs office was soon unified with the Reich, which led to the reorganization of foreign trade and export to Germany. The Czechoslovak crown was anchored to the German mark, and its circulation became more and more difficult. The "supreme authority of prices and values" instituted price and wage controls.[93] Meanwhile, the confiscation of state participations in large industrial enterprises and of Jewish citizens' property caused Czech capital to drop from 84 to 8 percent, while German capital grew from 0.2 to 62 percent.[94]

The militarization of the economy considerably increased the importance of heavy industry at the expense of light industry, craftwork, and building, and led to a redistribution of the workforce: by March 1945, 54 percent of industrial workers were employed in metalworking, 12 percent in other branches of heavy industry, 7 percent in construction, and less than 26 percent in light industry.[95] At the same time, the wartime effort and deportation of skilled personnel in the period of terror following the attempt on Reichsprotektor Reinhard Heydrich altered the composition of the workforce: women, young people, and workers from other professional and social levels flowed into the factories. All this, along with disastrous dietary conditions and actions of resistance and sabotage, caused the collapse of industrial productivity.[96]

The period of occupation introduced elements that produced a sharp break with the former Czechoslovak industrial tradition. It is true that the process of industrial rationalization and concentration begun in 1918 along Taylorist lines ended in these years, especially in the mechanical engineering industry. But this process took on forms foreign to and incompatible with both the MAP experience and Škoda management's attempts to reconcile the native craftsmanship tradition with elements borrowed from the American model.[97] Now principles of scientific management were applied in a context adverse to the democracy of the First Republic, which had been so important in determining their spread and acceptance. Moreover, their implementation was part of a war economy managed from abroad and aiming at the subjugation, or even annihilation, of the national native component. Divested of its ethical dimension in the harsh managerial techniques and the piecework systems that it used, the organization of work showed its alienating nature, not only regarding workers' creativity but also, and above all, their national and ethnic pride. It comes as no surprise, therefore, that the hostility of the population towards a system felt to be a symbol of their exploitation made it necessary to find new models of organization and management in the reconstruction period.[98]

From 1938 on, MAP's initiatives were reduced almost exclusively to combatting waste, in collaboration with the National Committee of Bohemia and Moravia for the Scientific Organization of Work.[99]

As far as Škoda was concerned, on December 23, 1938, the French company Schneider had sold its shares in Škoda to the weapons manufacturer, Zbrojovka of Brno.[100] On August 29, 1939, Škoda became part of the Reichswerke-Hermann-Göring A.G.[101] From 1940 onwards, the new German management adapted the entire production of Mladá Boleslav to the wartime needs of the Reich. Škoda changed the organization of work and with the help of private consultants introduced the method devised by the *Reichsausschuss für Arbeitsstudien* (REFA) [Reich Committee for Labor Studies]. Automobile production was gradually reduced, while truck production rose in 1939, only to be reduced once again in 1940, when the factory took on wartime orders. Civil production was converted firstly to manufacturing driving shafts and gears for Daimler-Benz and then to building aircraft wings, aeronautic equipment, artillery cartridges, anti-tank grenades, cannon parts, tank chassis, and military vehicles.[102] In 1942 Škoda was taken over by Zbrojovka Brno, giving birth to the Waffen Union Škoda-Brunn GmbH, whose administration was in Berlin.

Meanwhile, weapons production had grown rapidly, so much so that it employed almost 15,000 workers in a continuous cycle of three eight-hour shifts.[103] The highest level of exploitation came in 1944, when automobile production fell to a mere thirty-five units. The greater number of employees made necessary by the introduction of aeronautic production was produced by governmental decree in 1941, which made work compulsory for housewives and recent high-school graduates. In the following year, working hours, which until then had been formally limited to forty-eight hours a week, were lengthened to fifty-seven hours with only one daily break of not more than thirty minutes. This also applied to women and to children over sixteen, who even had to work night and Sunday shifts.[104] In March 1944 the twelve-hour shift was established.[105] In addition, from 1940 until 1944 there was a continuous deportation of skilled workers to German factories and work-camps and of traitors and saboteurs, some of whom were involved in the attempt to kill Heindrych, to Terezín, Auschwitz, Mauthausen, and Buchenwald.[106]

In 1945 the Mladá Boleslav plant was hit by nine Luftwaffe bombs. The shops, warehouses, and assembly departments suffered enormous damage. However, a considerable part of the machinery had been sequestered during the German withdrawal. After so many years of occupation, destruction caused by bombing, and the poverty of the workers, starting up a new production program was going to be difficult indeed.

Chapter 2 American Assembly Line and Czechoslovak Road to Socialism, 1945-48

Post-War Automobile Industry

In the years following the Second World War, the debate on new American production principles and techniques grew more heated, and the political stakes in economic reconstruction were higher than ever. On April 4, 1945, one month before Prague was liberated by the Red Army, representatives of the four Czech and the two main Slovak parties – *Komunistická strana Československa*, KSČ [Communist Party of Czechoslovakia], *Česká strana národně sociální*, ČNSP [Czech National Socialist Party], *Československá strana lidová*, ČSL [the People's Party], *Československá sociální demokracie*, ČSDP [Czechoslovak Social Democratic Party], along with *Komunistická strana Slovenska*, KSS [Communist Party of Slovakia] and *Demokratická strana* [Slovak Democratic Party] – announced the formation of the new National Front Government. They also stated the guidelines for reconstruction, called the Košice Program because of the Slovak city where it was presented.[1]

Despite the relatively moderate line followed by the National Front Government, we should not underestimate the influence of the Communist Party, which controlled the most important ministries. As Bradley Abrams has recently noted, in Czechoslovakia at that time there was a typically deep and widespread desire to break with the institutions of the First Republic and rebuild the country's international alliances, which had shown themselves incapable of protecting Czechoslovak sovereignty.[2] In a climate of fierce anti-German and anti-Hungarian nationalism and widespread mistrust of western European powers, Klement Gottwald was able to present the Communist Party not only as the representative of the working-class movement and the resistance, but also as the heir to national traditions and the socio-political message of the anti-Bolshevik President, Masaryk. At the same time, Gottwald freed himself from the inheritance of the interwar political institutions that still weighed on the other parties of the coalition.[3] The treaty of alliance with the Soviet Union, signed in 1943, continued to guarantee the security of the Czechoslovak state and, together with the Red Army's withdrawal immediately after the cessation of hostilities, created a particularly fertile terrain for the "Czechoslovak road to socialism" that Gottwald would launch in 1947.

Between May and October of 1945, President Eduard Beneš promulgated his controversial Ninety-Eight Decrees. They included the expulsion of the German minority (one-fifth of Czechoslovakia's population), the confiscation of Germans' and Hungarians' property along with that of collaborators, and the nationalization of key industries. With the Decree of October 24 (100/1945), 2,000 enterprises, representing roughly 60 percent of Czechoslovak industry, were nationalized and later combined according to sector or region into 250 nationalized enterprises [*národní podnik*], each with about 2,500 employees. The government, supported by the Economic Council – an advisory board with representatives of the various parties – created central directorates [*generalní ředitelství*] responsible for coordinating the enterprises in each industrial sector and managing the distribution of their output.[4]

On October 25, 1946, one year after the nationalization decrees, the National Front Government approved the Two-Year Plan, which clearly adhered to the political line of "socializing democracy" agreed on by the heterogeneous forces of the coalition.[5] The main goal of the Two-Year Plan was reconstruction and the recovery of the production levels and living standards enjoyed by the population in 1937. According to writer Alice Teichová, the plan aimed to set macro-objectives that left abundant space for the workings of the market, with a mixed system of small to medium-sized private firms together with nationalized strategic enterprises.[6]

The Two-Year Plan envisioned massive investments and a decisive intervention in the mechanization and rationalization of factories and of production in general.[7] It was thought that loans and international aid, available through the UNRRA (United Nations Rehabilitations and Reconstruction Administration) in the years immediately following the war, would at least partly finance these interventions.[8] In reality, this aid was suspended in September 1946 and, disappointingly for Czechoslovakia's expectations, it was not substituted by loans for reconstruction, probably because of growing American intolerance of the Czechoslovak government's pro-Soviet stance. Alice Teichová points out that in those years, the attitude of the American administration towards Czechoslovakia was vacillating and unclear. On the one hand, Washington did not want to encourage the National Front Government to strengthen its relations with the Soviet Union; on the other hand, it was afraid of furnishing aid that might somehow increase the country's industrial potential.[9] The turning point came in July 1947, when Czechoslovakia was invited to participate in the Paris Conference, which heralded the launch of the Marshall Plan. A telegraph from Gottwald, who was in Moscow, blocked the Czechoslovak delegation just as it was about to leave for Paris, thereby marking the end of all uncertainty as to Czechoslovakia's position in a divided Europe. Only a few months later, in February 1948, a serious institutional and political crisis led to the dissolution of the National Front Government and takeover by the Communist Party.[10]

The debate on the future of the automobile industry took place in a political climate marked by the desire to break with the institutions of the First Republic, a strong sense of national insecurity, and the search for solid material bases on which the Republic could be founded. Because of the inherent strategic value of their products, the largest mechanical and engineering companies were among the first to be nationalized. The former Škoda concern was nationalized on March 7, 1946, and divided into various independent enterprises: *Škodovy závody* [Škoda Works] in Plzeň kept the production of machine tools, heavy engineering and military goods; *Spojené závody pomocného automobilového a leteckého průmyslu, národní podnik*, PAL [Joint Auxiliary Works for the Automobile and Aeronautic Industry] inherited the auxiliary plants producing small electrical engines, electrical accessories, starters, and brakes; *Letecké závody, národní podnik,* Avia [Areonautic Works National Enterprise] took over aeronautic manufacture. The former ASAP in Mladá Boleslav kept the plants for automobile production and from then on was called *Automobilovè závody, národní podnik,* AZNP [Automobile Works National Enterprise], but its products kept the Škoda brand name. AZNP took over the plants of the previously independent automobile enterprises, Jawa of Kvasiny and Vrchlabí.[11] All these new enterprises depended on the authority of a central directorate, *Československé závody kovodělné* a *strojírenské, národní podnik*, ČZKS [Czechoslovak Engineering and Steel Production Works], with headquarters in Prague.

ČZKS coordinated various enterprises and production units in a number of related manufacturing sectors: steel production, heavy and light engineering, motor vehicles, electro- mechanics, and machine tools. The board of directors of ČZKS and the Ministry nominated the directors of the production units and some members of the management board, while one-third of the board was elected by workers' representatives. The nationalized enterprises were not formally part of the state administration, but they had the same legal status as joint-stock companies, with their capital entirely state-owned. They thus kept a certain margin of independence and had their own budgets.[12]

Table 2.1 – Czechoslovak Motor Vehicle Production Figures by Manufacturer and Model in 1939

	Automobiles		Commercial vehicles		
Manufacturer	**Models**	**Units produced**	**Models**	**Units produced**	**Total**
Aero	Aero 30	551			749
	Aero 50	198			
Jawa	Jawa 1	786			786
Praga	Praga Piccolo	1,198	Praga LN	163	
	Praga Lady	731	Praga AN	56	
	Praga Alfa	78	Praga RN	476	
	Praga Golden	8	Praga RND	181	
			Praga SND	132	
			Praga ND	142	
			Praga RV	15	3,180
Škoda	Popular 995	761	Škoda 104	97	
	Popular 1101	3,842	Škoda 206	121	
	Rapid	992	Škoda 254	121	
	Favorit	8	Škoda 266	5	
	Superb	71	Škoda 304	13	
			Škoda 404	99	
			Škoda 506	18	
			Škoda 606	215	6,363
Tatra	Tatra 57	1,707	Tatra 92	506	
	Tatra 75	104	Tatra 27	236	
	Tatra 97	282	Tatra 24/67	18	
	Tatra 87	185	Tatra 85	67	3,230
	Tatra 52	74	Tatra 85/91	51	
Walter			Walter PN	542	542
Total	17	11,576	21	3,274	14,850

Source: Škoda-Auto Historical Archive (Aša), Mladá Boleslav, Automobilové závody, národní podnik (AZNP), b. 34, Český Státní statistický úřad, zprávy a rozbory, rok 1970, *Rozvoj automobilového průmyslu v ČSR*, tab. 12.

An American Expert for the Czechoslovak '*People's Car*'

Compared to 1937, prospects for the automobile industry had completely changed. AZNP had taken on a sectorial and national dimension and was now destined for the horizons and potential of big business.[13] In 1946 the general manager of ČZKS, František Fabinger, promoted a project to restructure the Czechoslovak automobile industry with the help of an American engineer, Alexander Taub. His consultancy began on June 10, 1946, and ended after February 1948. It is documented in three reports written by Taub himself and notes taken by the Czechoslovak engineers responsible for production departments that cooperated with him.[14] How this collaboration came about is still not clear, as the documents do not reveal any

Table 2.2 – Czechoslovak Motor Vehicle Production Figures by Manufacturer and Model in 1948

	Automobiles		Commercial vehicles		
Manufacturer	**Models**	**Units produced**	**Models**	**Units produced**	**Total**
Aero	Minor	4,430	Aero 150	930	5,360
Jawa	Jawa 1	786			786
Praga			Praga RN Praga RND Praga ND	1,900 995 230	3,125
Škoda	Škoda 1101 Superb	13,907 52	Škoda 706	1,384	15,343
Tatra	Tatra 57 Tatra 87 Tatra 600	815 700 90	Tatra 111 Tatra 114 Tatra 115	345 99 70	2,119
Walter			Walter PN	542	542
Total	6	19,994	8	5,953	25,947

Source: Škoda-Auto Historical Archive (Aša), Mladá Boleslav, Automobilové závody, národní podnik (AZNP), b. 34, Český statní uřad, zprávy a rozbory, rok 1970, *Rozvoj automobilového průmyslu v ČSR*, tab. 13.

direct involvement of the American authorities. It is likely that the first contact was made thanks to Jaroslav Frei, ex-director of Jawa, who was responsible to Fabinger for developing a program for the automotive industry and who visited the U.S. in June 1946.[15]

Taub's consultancy in Czechoslovakia demonstrates that at least in the years immediately after the war, the fervent enthusiasm for the post-war reconstruction of Europe shared by American institutions and a multitude of entrepreneurs and adventurers was not in the least dampened by the uncertain destiny of central Europe.[16] The scarce biographical information available concerning Alexander Taub shows him to be an intriguing personage. Born in England, he worked as a designer for General Motors in the 1920s. He then became a consultant for the Office of Production Management (OPM) and the Office of Emergency Management (OEM), with the title of Chief Engineer for the Engineering Service of the Foreign Economic Administration (FEA) and the War Production Board. These agencies were in charge – though with continually changing responsibilities – of coordinating military production and provisioning under Franklin Delano Roosevelt.[17]

Apart from being a man who clearly enjoyed the favor of the Roosevelt administration, Taub was an ambiguous figure: in the last years of the war, he was in charge of producing a ten-year plan for the development of the Brazilian economy and industry; then in 1945, Chiang Kai-Shek engaged him to draw up a five-year plan for China's reconstruction, with the assistance of Chinese engineers.[18] Once back from the Orient, he founded his own consultancy company, Taub and

Associates; in 1946 he tried to interest the American administration in a plan for European post-war reconstruction based on the expansion of Polish industry as an anti-German measure, to be funded by the Import-Export Bank.[19]

Taub's history and the contents of his reports reveal an unusual ability to grasp the nuances of local politics and, above all, his unshakable faith in the "Politics of Productivity".[20] In his first report of September 1946, entitled *A People's Technology*, Taub started from the premise that it was with the automobile industry, the sector that symbolized mass production and the well-being associated with it, that the reconstruction of the Czechoslovak economy should start.[21] To this end, it was necessary to discover both the weak points of the local industry and the technological innovations that could best be adopted at AZNP. According to Taub, the new measures should take two directions: reducing production costs by finding a "people's technology" and planning a "people's car".[22]

Taub began his long first report by referring to the "root of the industrial problem", that is, the capacity to produce steel at a reasonable price. He wrote: "Live steel is needed, steel that will raise the standard of living, that will move goods, that will make things to make things with – and that means good steel at fair prices available freely for such living enterprise."[23]

Importing from the U.S. would allow the Czechoslovak industry to take the most urgent measures, namely the supply of motor vehicles and the mechanization of coal mines.[24] For the automobile industry to be competitive, a car had to cost no more than ten times the value of the basic raw material (steel), as in the U.S., and not twenty times more, as in Czechoslovakia in the summer of 1946. Also to be taken into consideration was the fact that American salaries were roughly double those of Czechoslovak workers. It was therefore imperative to reduce all waste in the manufacturing process: "Waste includes misuse of manpower, loss of machine-time by heavy removal of rough material, too much manual work and too expensive types of material."[25]

Taub complained about the lack of awareness of the fact that "waste generates costs" and about the absence in the organization of a specialist whose job it was to minimize the quantity of material kept in storage while awaiting finishing. In contrast, he acknowledged that the Czechoslovak technicians were on the whole familiar with automatic manufacturing methods and even paid excessive attention to laboratory measurements, for which they possessed equipment that, in his opinion, would suffice for the development of this sector for a decade. However, this competence and machinery were rarely used on the lines or in quality controls, and considering the limited volumes of production in 1946, their presence seemed purely 'academic' to him.[26]

The goal Taub presented to Czechoslovak managers and engineers was to enter the European automobile market, taking advantage of the weakness of the Germans and the impossibility for France and the U.K. to get back to pre-war levels of production in a short time.[27] It was a question of reaching the critical threshold of 200 cars a day in every plant, to be able to profit from the advantages of mass production, and to produce an inexpensive, low-powered utility car that would be affordable for a large number of people. It could expand the currently limited domestic market. This market had to be able to absorb at least 125,000 cars a year, and a car must not cost more than twenty-three thousand crowns, the equivalent of six months' wages for an average worker (according to Taub, the normal car price in the U.S. represented four months' salary for the ordinary worker).[28]

The idea of a "people's technology" was the thread running through Taub's reports. He seemed perfectly aware of the delicate political situation in post-war Czechoslovakia, knowingly alternating references to the "people's democracy" with ones to the "people's car". They were mixed in with references to "national undertakings", the core idea in the Košice Program's call for a united effort on the part of Czechoslovak technicians to a more organic vision of industrial development.

> The program is aimed at Technology for the people, a foundation stone on which a courageous country will endeavor to build a better living for its people and a better economic position for the nation. This requires a new type of management [...] since the men who will build this industry must be conscious of a greater responsibility than in private industry. [...] They must know their job as never before, and without being stubborn, hold out for what they believe is right.[29]

Beyond the renewal and the competitiveness of the automotive industry, it is clear that Taub's proposals were actually proposals for the reorganization of the entire national economy. They contained features similar to the development projects that Taub himself had designed for Brazil and China. Taub was aware that reliance on planning and on nationalization would face a great deal of criticism. But he also felt that the national welfare produced by planning would gradually mitigate negative judgments and cancel mental reservations.[30]

Along with obvious references to the current political debate in the U.S., Taub and his collaborators' convincing pages also contained astute comments about post-war Czechoslovakia's most critical political issues and the widespread anti-German sentiment. For example, commenting on the Czech technicians' uncertainty and bewilderment, Taub wrote:

> We appreciate that for centuries the Germans were masters and [...], teach that only they can be masters. This type of mental oppression can be used as an excuse for an over-modest attitude. But when it has been "live or die", the Czechs have never faltered and certainly we have recently met here many men who have what it takes. After all it is doubtful if a greater industrial master ever lived than Tomas Bat'a, unless it was Henry Ford.[31]

Taub concluded by recognizing the know-how demonstrated by Czechoslovak technicians in every phase of the production cycle, from metallurgy to the construction of machine tools. This know-how was the fruit of long years of experience in manufacturing vehicles in the Czech lands. [32] However, they still had to learn to abandon an attitude more suited to "organized craftsmanship" than to the more advanced mode of organizing production represented by the "economics of volume of production".[33] To that end, Taub proposed sending selected technicians to the U.S. to visit the major automobile manufacturing plants.[34] Taub stressed the need to train modern assistants to help the older managers. He intended to provide them with some knowledge of plant operations and equipment in countries "whose manufacture was in advance of Czechoslovakia". Finally, there was the question of the over-abundance of specialized manpower. Taub stressed the noteworthy competence of skilled Czech workers.[35] According to Taub, most of them could be transferred to manufacturing machine tools or prototypes. In fact, to employ such highly skilled workers on machines would mean their de-skilling.[36]

To accompany these changes, Taub thought to form a team of "worker trainers" and send them to America. There they would take courses given by the Jam Handy company, which provided visual aids, and with the Chan Dooly Group, which had been in charge of workers' training in the U.S. during the war. To fill the places vacated by the transfer of skilled workers to the machine tool industry, the mechanization of agriculture and the reform of the mining industry were necessary to make new unskilled laborers available.

The American model of mass production was a reference point not only for training managers and the layout and organization of plants, but also for the structural and organizational transformations that had led the American automobile industry to centralize, specialize, and standardize the production of cars and components.[37] Taub stressed the weakness and lack of competitiveness of plants with low production volumes and small dimensions.

These remarks obviously referred to the current Czechoslovak debate on *malost* (small dimensions): the fact that this was a small country with small businesses and relatively limited volumes of production had, according to some, determined the political fragility of the First Republic and made its defense all but impossible.

Czechoslovakia came out of the Second World War even more certain than in the period after the First World War that the political survival of a state of Czechs and Slovaks depended on very concrete things: a powerful army and an efficient and productive industrial sector. Political disappointment with the European powers, seen as Hitler's accomplices, was associated with admiration for the way in which the two great powers to which the Czechoslovak state owed its newfound independence had managed the war effort.[38]

In this context, Taub's invitation to centralize, standardize, and rationalize automobile production, by eliminating the industrial fragmentation typical of the private sector before the war and the tendency to internalize all auxiliary production, could not go unheard. Nationalization of plants and the creation of AZNP would open the way to "[...] operate an entire sector as if it were a single productive unit", eliminating waste and uneconomic practices.[39]

The model referred to in this case was not Ford, but General Motors, a sort of "mosaic of mid-sized and small businesses" supported by producers specializing in component parts that functioned, in turn, as "central plants".[40] Undoubtedly, AZNP, being a union of nationalized factories, was well-suited to this type of organizational structure, and the insistence on coordinating production in units that were separate and geographically distant corresponded to the interests of a balanced national economy, offering solutions to the still-unsolved problem of the industrialization of Slovakia. [41]

Taub's project envisioned a central engine plant for the national automotive industry (possibly three: one for motorcycle engines, one for cars, and a third for commercial vehicles), which would have its own foundries and metallurgy department as well as a development department. The other plants envisioned were: one for transmission and axle differential; one specialized in bodies and sheet metal, which would also include facilities for research on paints, surface and finishing treatments, and design; a series of plants in various parts of the country for smaller-size accessories (carburetors, electrical parts, breaks and gear boxes, tires, glass, and small parts – in this case, obtaining the highest degree of interchangeability of parts was the main objective); an assembly plant, which was the most important of all, for coordination and quality control. A large proving ground with a central research and development department completed the picture. To coordinate it all, Taub envisioned an Automotive Industrial Council (AIC) composed of one hundred individuals with well-defined lines of authority, who would guide and counsel on all matters affecting the growth of the industry and its programming and planning. The AIC motto was to be: "*skutečnost je příležitost*" [reality is opportunity].[42] The AIC would have been a complex body assisted by a commercial department.[43]

Its scheme included a director at the top representing the General Manager of the Metal Industry assisted by a strong executive secretary. Under the Director and executive secretary were to be the deputy directors representing automobiles, trucks, motorcycles and tractors. Each deputy would have been responsible for the implementation of the national program in his branch of industry. Metallurgy, research, standards, and central manufacturing plants would have been common to all branches, and a deputy director had to be provided for guidance of the activities whose products or services were common to all.[44]

Plants and People's Technologies

Taub's report already presented a hypothesis for the allocation of labor in the factories coordinated by ČZKS, based on the actual conditions of the automotive plants in the country.[45]

The plant for producing differentials and gear boxes common to trucks, automobiles, and tractors could be in Libeň, an historic working-class neighborhood of Prague, which produced truck engines, diesel and gasoline engines, and automobile engines, although it was actually a differential and transmission plant. It was in fact well-equipped to produce gears and gear boxes, and during the war it had produced the Wilson gear box. Over 70 percent of its capacity had suffered war damage and needed to be rebuilt. According to Taub, it could therefore be reinstated as a plant specialized in its original production line, so as to "concentrate the present knowledge and new processes to come".[46]

The Vysočany plant had produced Junker aircraft engines during the war. In 1946 it was assembling truck engines and making truck cabs. According to Taub, after considerable updating of methods and machinery which would include the introduction of presses and die shops for all truck sheet metal and also of welding equipment for assembly, this could become the central truck body plant.

The huge Čakovice plant was used by the Germans to manufacture aircraft engines. After 1945 it produced injection pumps for diesel engines as well as components and the assemblage of seven-ton trucks. Taub felt it could be used to produce truck and tractor gasoline and diesel engines. This would require building a foundry and better utilization of machinery.

As far as automobiles were concerned, Taub thought that the complete standard passenger car could be made at Mladá Boleslav, where the Š1101, ex-Tudor, was still in production. The plant had been bombed in May 1945 and had suffered damage to over 15 percent of its facilities. What is more, the body shop, once called

America, was totally unsatisfactory because it occupied several floors. Taub was very critical of its production methods. In his opinion: "Machine shop philosophy is slow. Too many single operations, too much material removed. Design of parts is expensive, making assembly of chassis costly."

He thought it would be necessary to redesign the whole plant and build from scratch the body shop, foundry, sheet metal shop and assembly lines. Only then could Boleslav produce the required three hundred cars per day. Finally, there was the Tatra factory at Kopřivnice, which Taub considered to be the most important, not so much the plant itself or its equipment as its very high caliber technical personnel. Here they were preparing to produce six thousand Tatra 87 cars. They made heavy trucks with twelve-cylinder, air-cooled diesel engines as well as freight cars. They were also proposing to build locomotives.

The Slovakian plant at Povážká Bystrica would produce 100 cc motorcycles and other important parts, including those used for cars. As in the case of the Tatra, Taub praised the high caliber competence of this plant's technical personnel, both for their experience in the manufacture of machine tools and for the methods used in automatic inspection.

Motorcycle manufacturing represented another of the strengths of the Czechoslovak automotive industry, and Taub envisioned a total production of 125,000 units per year divided into five models. At Brno in Moravia, which produced tractors, they would manufacture 250 cc and 350 cc motorcycles, while Strakonice, which was already partially equipped with automatic machines, would make 175 cc motorcycles. Finally, the Jawa plants in Prague and Brodec would need to be converted to manufacture truck accessories.

As regards automobile production, Taub envisioned four models with a high degree of interchangeability of parts, for a total of 125,000 cars per year: a two-cylinder, 445 cc car weighing 500 kg, a new version of the Jawa CZX; a four-cylinder car weighing 690 kg, with an 890 cc engine; a six-cylinder 1,350 cc car weighing 750 kg to be made at the Škoda factory in Mladá Boleslav; and finally, an eight-cylinder car with a 2,500 cc engine weighing one ton to be made at Kopřivnice. The benchmark for comparing the ratio between manufacturing costs and automobile weight for the mid-size 690 kg car was the English Vauxhall ten hp.

In a second, undated report written in Czech by Alois Hrdlička, the director of AZNP, the picture seemed somewhat different.[47] They had hit on the idea of an ultra-light car. Three models now remained: a four-passenger, 1,200 cc car weighing 700-750 kg (75,000); a five-passenger, 1,800 cc car weighing 900-950 kg (25,000); and a six-passenger car with a 2,500 cc engine weighing 1,050-1,200 kg manufactured by Tatra (6,000-10,000 cars per year). The authors questioned the choice of vehicles and the site where they should be manufactured, concluding that none of the existing factories could achieve a production of 50,000 cars per year with only one shift, or possibly achieve 100,000 cars annually with two shifts. It would therefore be better to make the four- and five-passenger cars in a new factory complex. Before they planned the new body and die cast shops and the foundry, Taub advised the Czechoslovaks to visit Detroit, where they would find lists of the required equipment as well as technical specifications.[48] Meanwhile, they designed two prototypes for standard passenger cars to be ready by October 1947: one at Mladá Boleslav for a rear-drive, 750 kg car and the other a front-wheel drive.[49]

Taub's report reveals the tensions existing among the directors of the factories that had produced cars in the pre-war period, especially Jawa, Škoda and Tatra, none of whom intended to give up their automobile production lines. In this second report – perhaps thanks to Hrdlička's contribution – the Jawa brand, earmarked as the "light car" and representing the peak of Czechoslovak industry, now seemed to be discarded, along with the idea of manufacturing a light car. It is important to note that in 1947 already and through the early 1950s, all reports, Taub's as well as those written by Czech specialists (as we shall see later), show concern for the deep sense of belonging to their factory and their brand shared by technicians, workers, trade unions and local party cells. This defense of local interests took on the value of an identity, and it never ceased to be a problem for planners and those attempting to rationalize the system.

Taub's proposal was realized in the third and final project for a new automobile manufacturing plant.[50] The new factory would be situated in a location with good river and rail connections and would be made up of three main sections – mechanical works, body and assembly shops, and foundry. Production in this new factory was to reach 50,000 vehicles a year, with only one eight-hour shift.

The first novel aspect of the Mladá Boleslav factory was the storage of materials required for manufacturing, both rough materials and pieces bought already finished and ready for assembly. As in American factories, no large central warehouse was constructed, and preference was given to a solution that would allow for stockpiles of material needed for fifteen days to be kept near the lines inside the shops. It was also necessary to increase and improve routine controls.[51]

Transportation was also facilitated by constructing loading-ramps for trucks in the various sections, to allow them to unload materials close to the site where they would be used. For this reason, the assembly shop had to be situated half-way between the body shop and the mechanical shop. In Taub's plan, the body manufacture shop was connected to the pressing shop. It included a warehouse for materials, a section equipped with large presses, another section with small and medium-sized presses, a pre-assembly area, an assembly area, and finally the bonderizing, painting and finishing shops. Careful checks would also eliminate the majority of manual work in the assembly phase, which was considered to be the chief reason for high production costs. In addition, the vicinity of the press shop and the availability of presses would make it possible to substitute or repair defective pieces very quickly.

As regards the mechanical shop, Taub envisioned the utilization of tool machines manufactured in Czechoslovakia, especially drills. Finally, various oval or circle-shaped lines were planned for the manufacture of pieces requiring complex operations. In this shop, too, there would have to be assiduous controls of raw materials to guarantee precision and minimum tolerance. The investment envisioned was huge: twenty-eight million dollars, not counting the machine tools. Nonetheless, Taub felt it could be reduced to fifteen million dollars for structures and five million for machinery. He hoped to be able to produce the machinery in Czechoslovakia and purchase only the machines, pieces, and designs that were strictly necessary. The success of this project depended on the creation of research laboratories and testing facilities.

In its essential outline, Taub's project represented a simple, well-formulated synthesis of the achievements of the American automobile industry in the first half of the twentieth century, from the Model T to the multi-divisional structure of General Motors.[52] It would have to be carried out in stages: the reform and rationalization of the mining industry, which would guarantee a greater availability of raw materials at reasonable cost; and then agrarian reform and the mechanization of agriculture, which would make low-cost labor available to industry, as in Detroit in the 1920s. The skilled workforce that was so abundant in Czechoslovakia would find employment better suited to its capacities and expectations in the production of prototypes and machine tools. The admittedly well-prepared Czechoslovak management would be partnered by new assistants, specifically trained according to the rules of what seemed more and more a sort of mystique.[53] Finally, the move towards mechanization and automation of plants was required to reduce wastage of materials and workforce that was not compatible with an economy of production that could guarantee high wages to the workers.

Taub's consultancy anticipated and had much in common with the productivity missions that were to follow the Marshall Plan, which have been amply documented by historians. In their many technical assistance programs, the Americans who traveled to Europe in the years immediately following the war shared a particular way of relating to the European reality, which has been described in the abundant, finely nuanced literature on the subject.[54]

Yet what distinguishes Taub's consultancy from other reports by visiting businessmen, engineers, and consultants from across the ocean was the fact that in Czechoslovakia, the path to modernization indicated by Taub, while closely retracing the steps of American industrial development, was to be made possible by planning, by nationalization, and by "national undertakings", all elements essential to Gottwald's political project. In 1947 the American model and the Czechoslovak path to socialism were to some extent interdependent. This path guaranteed the institutional *tabula rasa* that would allow the country to reformulate its institutions. This was a necessary premise for redesigning industrial development along the lines of the American model and also provided the national rhetoric needed to overcome possible resistance. On the other hand, American technology and organization guaranteed the efficiency and productivity essential for the survival of the "national path". It was a vicious circle that could have allowed Czechoslovakia to remain an "island of democracy" between what was being delineated, to quote Paul Hoffman, chief administrator of the Marshall Plan, as the confrontation between the "Communist Party line" and the "American assembly line".[55] But, as we shall see later, it could also have imitated the "American road" chosen by Lenin to modernize Russia. In 1947 it was difficult indeed to trace the line separating the Soviet industrial model from the American model.

As the network of references included in his reports shows, Taub was probably aware of this. But it is impossible to know the real aim of his mission: were his own personal interests involved? or was he taking orders from the American Department of Commerce or some other institution? or was it simply that he really did believe in what Czechoslovakia was trying to do? Hindsight, however, makes it clear that the "conviction that class conflict could be overcome through national unity and well-being guaranteed by efficiency, scientific management and planning" in people's democracies would favor not so much America's interests as those of the Communist Party's propaganda and the Soviets.[56]

Yet rather than the motives that moved Taub to visit Czechoslovakia for over two years, it is perhaps more interesting to understand how his message, so strongly rooted in the "politics of productivity", was received in the bordering countries generally unfairly excluded from the lengthy discussions on Americanization of parts of Europe included in the Marshall Plan. Extending this discussion to the bordering countries identifies two forms of "American engagement" concerning the choices needed for rebuilding the automobile industry: one linked to the influence of Taub's consultancy, the other mediated by the Soviet experience.[57] Realizing how these two ways of interpreting mass production, so different despite their common origin, were superimposed in Czechoslovakia is key to our understanding the evolution of this country's industrial culture.

Technical Pilgrimages to America

According to an agreement made with Taub, in those same years a group of technicians from AZNP Škoda set out for America. Some of them had held positions of responsibility in the second half of the 1930s and would remain with AZNP until the 1960s.[58] Although the former director of Škoda Auto, engineer Karel Hrdlička, who refused to become a member of the Communist Party, was unjustly accused of collaborating and forced to earn a living with technical translations, other specialists continued to occupy roles of responsibility. Some of them played a very important part in the history of Czech automobile production, including Vladimír Matouš,[59] Zdeněk Kejval,[60] and František Žalud.[61]

These trips were meant to eliminate the technological and organizational gap that Taub had so strongly stressed, at least in theory. The Czechoslovak technicians visited the major American automobile companies as well as machine tool and component manufacturers. They also visited cultural institutions, museums, technical colleges, universities, and research centers. Their reports incorporated with Taub's project for rebuilding the Czechoslovak automobile sector help us to evaluate the Czechoslovak technicians' response to his proposals. During their visits, the technicians concentrated mainly on what Taub had identified as the weak points in Czechoslovak automobile production. What particularly stands out this time is an interest in technical features and the ways in which new machinery was utilized in overseas plants and in the progress achieved by the American industry thanks to the war. They were detailed reports, in some cases followed up by contacts inquiring about the possibility of supplying machinery.

As regards the use of machinery on the line, the technicians emphasized the new complexes of machines with huge production capacity, and they noted that the amount of special machinery had greatly increased from before the war.[62] The reports dedicated page after page to single-purpose machines mounted on automatic lines, which made it possible to carry out a series of operations without the worker having to maneuver the machine, except to turn it on and off. Once again attention was focused on technical devices that reduced effort and exertion on the part of the worker and at the same time made work more productive (for example, pneumatic tools that hung balanced beside the worker's hand). Another important aspect was adequate lighting in the shops, obtained by a combination of cold and mercury vapor neon light.

Compared with the pre-war period, considerable progress had been made in the manufacturing of surfaces; the technicians noted here the use of Superfinish[63] and the new Natco machines, the Healds, the Ex-Cell-O, presented as indispensable for up-to-date production.[64] Other things they admired were the speed and organization of assembly lines and the use of networks of teletypewriters.

However, the innovations that struck the Czechoslovak technicians the most involved the manufacturing of bodies, the use of equipment for moving and transport inside the factory, and the new ways of storing materials. The Americans' progress in body manufacture was a result of experimentation in the aeronautics industry. It had determined a notable superiority for the U.S., and the gap in this area compared with current Czechoslovak production was considered insurmountable, and not only because of the lack of machinery. The technicians emphasized the widespread use of the spot welding machine in body manufacture, other large welding machines for the various sheet parts, the so-called ultra-speed welding machines[65] supplied by Progressive Welder Co., and the different presses – mechanical supplied by Bliss & Clearing and hydraulic supplied by Hydraulik Press. Finally, they noted that the material used in America was top quality, especially the sheet metal, and this greatly helped technological sections and contributed substantially to the quality of the finished product.

The Czechoslovak technicians reacted not only to the machinery and factory organization but also to other novelties they saw in America. Vladimír Matouš, for example, highlighted the American automobile industry's tendency to gradually subordinate technological innovation to issues of profit and marketing strategies. The greater importance of Sloanist logics led to the production of a vast range of models based on the use of the same mechanical parts. The reproduction of models (deluxe and sports) whose differences lay (and still lie) in minute details of design and accessories was soon to become common in Europe as well.[66]

It was clear that AZNP would need to make gradual but significant investments ranging from improving existing plants to far more radical measures, including building and equipping a completely new factory. Matouš was a keen observer and showed an acute understanding of the American situation, market mechanisms, and trends in production.[67] His words further confirm the dedication and skill of the Czechoslovak technicians. At the same time this elderly engineer was distancing himself from the American model, as he wrote: the American technical solutions "represent only suggestions for our industry, and should not be taken literally. When we observe and when we buy, we have to study the spirit and the method of American practice".[68] And as regards projects for automobile construction:

> Even while taking into consideration our different needs, we can still learn American practice from many examples. First of all, this is work based on

> the search [...] for a series of details in construction and production, and in the use of materials for the chassis and the body. American planning also offers an example of what is required for long-term economic production and for the creation of an industry capable of prediction and economic orientation, and also the spirit of collaboration shown by American technicians in their work.[69]

Almost all these reports ended with the recommendation to purchase American materials and expressed serious concern about the destiny of the Czechoslovak national industry if it failed to modernize plants and machinery.

Despite their admiration for the productivity and modernity of American factories, the Czechoslovak technicians took a critical stance regarding some of the things they had seen in America. The general observations of the American social and political scene that were so abundant in the reports of the 1920s had disappeared. The American industrial relations system was totally ignored, despite the fact that the technicians frequently commented that they had not been able to visit factories because of strikes.[70]

This explicit and at times even flaunted indifference to social and disciplinary matters, working conditions, and labor-management relations had a strong impact on the Czech technicians' first encounter with scientific management, possibly due to the type of mission these technicians had been sent on – to compile a list of machinery to purchase for the factory that was going to replace *America* – as well as the Czechoslovak factories' urgent need for materials. But it could also be evidence that in the eyes of the technicians who went later, the American model seemed to be made up of different elements; the neutral ones they considered suitable for the Czechoslovak context were studied to the last detail, while others which they found less adaptable or even less desirable were disregarded as cultural or political peculiarities that had no place in what was to become the Czechoslovak route to socialism.

Moreover, the technicians were well aware of the enormous gap separating the actual conditions of Czechoslovak plants from Alexander Taub's vision and of the consequent necessity to find a Czechoslovak way to achieve mass production. Several accounts clearly show their awareness of the need to adapt the knowledge and techniques they had learned in the U.S. to the Czechoslovak situation in order to achieve European production levels.[71] In the course of 1947, other technicians had gone to France to visit Renault factories, to Italy to study body manufacturing procedures, and to Vauxhall in Great Britain.[72]

Frei wrote:

> We should not let ourselves be overawed by the enormous dimension of American production [...] or think we have to introduce everything they have: our friend Taub says: the grass is always greener on the other side of the fence, but only until we get close enough to notice the parts without grass that can't be seen at a distance [...]. The road that leads us out of the difficult situation of our automotive industry exists: finding how best to introduce American know-how in a Czechoslovak way, so as to arrive at a European model as soon as possible.[73]

The real problem, however, was how to define this Czechoslovak road and how in the end it could bring Czechoslovakia to be considered a European state to all intents and purposes. All the reports showed bitterness about the inadequacy of the national plants and dissatisfaction with the scarcity of machinery. Criticism of the incompetent bureaucrats responsible for the slowness of production went along with complaints about a lack of investment.[74]

However, it is difficult to find deep reflections on how to alter the American model so as to make it suitable. This may be the most interesting aspect emerging from an analysis of these reports: we can detect a profound gap, not so much in the way these technicians viewed America as in the way they saw the reality of production in their own country on their return. As in the 1920s, the political and intellectual debate in due course influenced the technical debate on managerial ideas and production procedures and how they were received. But unlike in the past, the theoretical speculations now found no barrier in the specifics of production and local markets or in the careful supervision of Klement who, along with Matouš, chose the machines one by one, pondering over their utility in the actual production cycle.

In this climate of industrial regeneration, almost everything had become possible. By 1947, little was left in Czechoslovakia of the multiplicity of suggestions and alternatives regarding procedures, models, theories, and suppliers of the multi-focused inter-war industrial culture that MAP, Klement, and Baťa had expressed. The technicians' reports were imbued with the ideological and technocratic aspects of the American model and its European version. Despite their insight and high level of dedication, they were not able to form a concrete alternative proposal to close the gap in resources of which they were acutely aware. On the eve of February 1948, the Czechoslovak road risked becoming little more than wishful talk inspired by nationalistic aspirations, rather than the result of a precise analysis of the reality of local production. In short, it was a container that had lost its contents one by one, and that revealed all its fragility when set against Stalinist certitudes.

Perhaps the only powerful glue was the trust expressed in the technical personnel and the high level of competence widely found in the Czechoslovak industry.[75] One of the things learned from the Americans was the use of competent, specialized personnel in every situation, without leaving anything to chance. What was needed was an appeal for the recognition of the role the technicians could play in reconstruction, thereby performing a task of great importance for the state.

> It is fundamental that, as in the U.S., work in the automobile industry be done by those who know it best. [...] If we apply all our knowledge and proceed according to the rules we are given by our technicians, we can find a solution for the difficult task we have in front of us today.[76]

Chapter 3 From American Dream to Soviet Management, 1948-51

First Five-Year Plan and Czechoslovak Motorization

The period that followed the end of the National Front government witnessed a gradual redirecting of Czechoslovak economic and political relations towards the Soviet Union. Between 1948 and 1949 the way was paved for a tenser international political situation and the division of Europe into two antagonistic blocks. The launch of the Marshall Plan, the Berlin crisis, the Atlantic Pact, the creation of Comecon (Council for Mutual Economic Assistance), and finally the Korean War marked progressive abandonment of the Czechoslovak road to socialism and the first phases of the Sovietization of the country.

In the course of 1949, the idea of possible collaboration with the West, considered indispensible for obtaining the technical support needed to modernize industrial plants, grew increasingly remote. After the Cominform (Communist Information Bureau) meeting of 1949, Czechoslovakia defined its alliance with Stalin's Soviet Union by condemning Yugoslavia and the "national road" to socialism and, from the economic standpoint, by signing the 1951 Czechoslovak-Soviet trade agreement.[1]

The First Five-Year Plan was drawn up in 1948; it became law in October of the same year and was modified in 1951. It was inspired by principles profoundly different from those underlying the Two-Year Plan (1947-48). The goal of a balanced development of the Czechoslovak economy had become secondary to that of reaching a higher rate of growth and investment through structural changes and the expansion of heavy industry, especially chemicals, and through the decided encouragement of Slovak industrialization.[2]

The plan was drawn up in the context of a fluid international situation. The original version, written on the eve of February 1948, was the fruit of a compromise between the requests of the Social Democratic Party, ČSDP, which upheld the need for balanced development, and those of the Communist Party, KSČ, which aimed at higher growth and investment rates. But after Gottwald's party took over power in February 1948, the latter aims prevailed. In 1951 the plan underwent a third revision: investments in the capital goods sector increased even more at the expense of investments in the production of consumer goods. Moreover, in the same year it was announced that it would not be possible to reach a 60 percent growth in the population›s living standards by the end of the Five-Year Plan, as promised when the plan was first published. Indeed, the year 1951 marked an important turning point, with an acceleration of the transfer of the Soviet system of industrial administration and management to Czechoslovakia. In the same year, the arrival in Prague of Soviet advisors coincided with a change in the internal equilibrium of the Communist Party, paving the way for the political trials of 1952-54.[3]

In the course of the First Five-Year Plan not only was the complete nationalization of all important economic activity carried out (95 percent of the Czechoslovak active population was now employed in the public sector), there was also a strong push towards industrial rationalization and concentration and the creation of large, multi-facility enterprises. At the same time, the strategic decision-making process was centralized, with the consequent transformation of enterprises that had until then enjoyed some degree of independence into mere units of production devoid of any authority concerning products, processes, salaries, customers or suppliers.

The composition of the organ responsible for formulating and carrying out the plan, *Státní úřad plánovací* (SÚP) [the State Planning Office], had already been partially modified in February 1948. This organ was transformed from a commission made up of technicians of different political orientations into a state planning office with ministerial functions, appointed by the government and the central committee of the Communist Party.[4] The ministries representing the various industrial sectors depended on the SÚP and coordinated the central directorates and nationalized enterprises.[5] The formulation of the plan and its execution were determining factors in transforming the Czechoslovak economy into a planned, Soviet-type economic system or, according to Janos Kornai's definition, into a "political economy of Communism". In a very short time they profoundly transformed Czechoslovakia's political and economic institutions under the pressure of what was by then commonly called *sovětské řizení* [the Soviet management system].

The organization of the automotive industry also underwent profound changes. With the First Five-Year Plan, the production of automobiles was penalized in the allocation of resources, as consumer good considered by the planners to be a useless luxury. Its marginalization, however, did not make it immune from continual interference in its organization, which aimed at carrying out the rationalization begun during World War II and continued after nationalization in 1946.

By government decree, on April 27, 1949, ČZKS was divided into three central directorates: *Československé závody těžkého strojírenství, národní podnik*, ČZTS [Czechoslovak heavy engineering works]; *Československé závody lehkého kovoprůmyslu, národní podnik*, ČZLK [Czechoslovak light engineering works], a national enterprise meant to coordinate production in light engineering; and *Československé závody automobilové a letecké, národní podnik*, ČZAL [Czechoslovak automotive and aviation industry, national enterprise]. ČZAL became the central administration responsible for the management of automobile, motorcycle and aircraft production.[6] It directly managed eighteen production units organized into four operational centers according to their production programs:

1. Production of automobiles and components (AZNP, Karosa, Tatra Kopřivnice);
2. Aircraft production (AVIA, Rudý Letov);
3. Motorcycle and bicycles and spare parts;
4. Repair shops, electrical components, brakes, and carburetors (PAL).[7]

The planners' ambition was to create a single multi-facility enterprise for each operational center by means of the amalgamation, suppression and reconversion of existing enterprises and production units. Coordinated and managed by the central directorate, ČZAL, the multi-facility enterprises would achieve capacity suitable for efficient production. As regards the possibility of mass production, the fact that these large enterprises were not only nationalized but also operated as monopolies on a national scale was considered an advantage.

From the point of view of its structure, ČZAL was very similar to ČZKS: there was a general director, to whom the central secretariat and seven departments were answerable (the department of planning, coordination and statistics; review and control of finances and accounts; center for operations and definition of production plans; cadres and political and social security department; technical sector for research and development; business finances, financial planning, budgeting and pricing; purchases and sales).[8] What was new compared to the period before 1948 were the greater restrictions on the autonomy of enterprises which lost their independent management and were deprived of a large part of their remaining decision-making powers.[9] Relations among the various national enterprises were still determined by contracts, but they could be modified from on high, by ČZAL. These contracts, which up to 1948 had still been partially governed by market mechanisms, were progressively emptied of all meaning, to the point that they became mere technical instruments that were in any case subject to vertical, hierarchical constraints.[10]

In 1951 a new reorganization eliminated ČZAL, and management of aircraft production became separate from land transport vehicles – *Československé závody na výrobu vozidel, národní podnik*, ČZVV [Czechoslovak motor vehicle production works, national enterprise]. A few months later the Ministry of Industry transferred and merged ČZVV into *Československé závody všeobecné kovovýroby, národní podnik,* ČZVK [Czechoslovak general engineering and steel production works], while the Ministry of Defense was in charge of the "militarization and Sovietization" of aircraft plants.[11] Increasing international political tension and the growing influence of Soviet advisors over the Czechoslovak central planning office also affected the production programs of the motor vehicle sector. In the months following the dismantling of ČZAL, division of work and production programs among the various automobile plants was modified to make room for military production. At the same time, the Soviet work management methods were introduced in the factories.[12]

Before this happened, in the first year of the Five-Year Plan, a heated discussion arose among ČZAL managers, factory directors, and local and central Party officials concerning the function of automobile production in that phase of building socialism. Certain articles and documents published in the journal of the Czechoslovak Automobile Club, *Svět motorů* [the World of Motors], enable us to understand the prospects for motorization in Czechoslovakia in 1949.

A few months after the launch of the First Five-Year Plan, discussions on the features of the automobile that could lead to Czechoslovak motorization alternated with declarations of the irrelevance of a policy favoring the spread of the automobile in Czechoslovakia. The latter emphasized the purely instrumental value of the automobile, seeing it as a "superfluous luxury" to which only limited resources should be dedicated.[13] Those who still supported the idea of producing a new car, exploring production programs and the future of the sector were referred to the plan: mass production of automobiles did not figure among the priorities of the First Five-Year Plan and to rethink this point was inconceivable.

Minutes of ČZAL meetings show that the new priorities for the transport sector were to expand the production of commercial vehicles, which could also serve for military use, and to give preference to models that could run in all weather conditions (this held for automobiles as well). Although the need to develop road transport was recognized and automotive production was considered an indispensable stimulus to develop other manufacturing sectors of the Czechoslovak economy, what dictated conditions to ČZAL were the requirements of a "motorized army" and the demand from foreign markets. In fact, ČZAL had to answer both to the Motokov, the general management responsible for the export of metal and mechanical engineering products, and to the Ministry of National Defense.[14]

Faced with centralized pressures barely concealing the existence of plans that cast doubt on the very survival of Czechoslovak automobile production, ČZAL technicians did not deny the validity of the plan nor the importance of its objectives. At the same time, however, they insisted on the necessity to implement the plan with flexibility and, above all, circumspection. This was dictated by the requirements of exports, but also by the hope of laying the groundwork for the future motorization of the country when the time was right.[15]

According to those responsible for the sector, the Party strategy suffered from certain contradictions that would make it very difficult to start up automobile production again when it was finally considered a priority and, in the long term, would also harm exports. Indeed, as the data in the appendix confirm, 75 percent of automobile production was destined for export to the area of the dollar; these proceeds were largely reinvested in manufactured products connected to heavy industry. The specialists noted that unless at least a part of the currency derived from the sale of cars was reinvested in modernizing facilities and in the expansion of production required to reduce costs, Czechoslovak products would soon not be competitive on the market. They pointed out that the rules of the game for exports, especially to countries with hard currencies, were different from those for the domestic market. On the international market, planners had to take account of the requirements of demand, and prices and products had to match supply. Such considerations cast a shadow on the radiant future of motorization envisioned by the planners.[16]

Leafing through the magazines of the period, we can see that though the possibility of limiting national automobile production to the minimum was taken into consideration, even in these years the line condemning motorization was not crossed in Czechoslovakia. On the contrary, the idea of proletarian motorization as opposed to pre-war bourgeois motorization was played up, and the planners presented the sacrifice of the automobile and motor vehicle industries as something temporary. Once socialism had been built and the risk of a new conflict had disappeared, they would tackle the issue of the motorization of Czechoslovak society and the mass production of cars.[17]

In fact, starting from 1951, the Czechoslovak Automobile Club became *Svaz pro spolupráci s armadou* [Union for Collaboration with the Army], and articles in *Auto* reframed the tasks of motorization as education for defense, road safety, and sport.[18] At the same time, socialist motorization was presented as motorization for the masses, in contrast to the bourgeois, élite motorization that had been a status symbol in Czechoslovakia between the two wars.[19]

Table 3.1 – Automobile Production at AZNP Mladá Boleslav Plant, 1946-55

Year	Model	Units produced
1946	Tudor Š1101	942
	Truck (2.5 t)	1,470
1947	Tudor Š1101	5,296
	Truck (2.5 t)	709
	Frame Š1101	918
	Superb OHV	28
1948	Tudor Š1101	10,186
	Semifinished Š1101	586
	Frame Š1101	2,752
	Superb OHV	103
1949	Tudor Š1101	4,469
	Cabr. Š1101	936
	Sedan Š1101	5,307
	Semifinished Š1101	1,400
	Frame Š1101	2,962
1950	Tudor Š1101	4,501
	Cabr. Š1101	998
	Sedan Š1101	6,524
	Semifinished Š1101	1,403
	SW Š1101	85
	Frame Š1101	3,676
1951	Tudor Š1102	3,602
	Sedan Š1102	5,504
	Semilavorato Š1102	850
	Frame Š1102	6,364
	Tatraplan T600	236
1952	Tatraplan T600	1,864
	Sedan Š1200	2,000
	Frame Š1200	10,143
1953	Sedan Š1200	2,600
	T805 military vehicle	2,098
	Frame T805	2,631
	Frame Š1200	2,007
1954	Š1200	2,933
	Frame Š1200	1,228
	T805	2,811
	Frame T805	37
1955	Š1200	8,001
	Š440	2,105
	Frame Š1200	4,422
	T805	13
	Š706	40

Source: Škoda-Auto Historical Archive (Aša), Mladá Boleslav, Automobilové závody, národní podnik (AZNP), b. 5, *Automobile Production at Mladá Boleslav Plant*, 1946-55, March 26, 1954.

Table 3.2 – Motor Vehicle Production (in units) per Year in Czechoslovakia, 1946-68

Year	Automobiles	Trucks	Buses	Total
1946	3,792	2,952		6,744
1947	9,372	5,237		14,609
1948	17,971	7,221	1,122	26,314
1949	20,769	5,733	1,021	27,523
1950	24,463	5,984	977	31,424
1951	17,064	11,530	988	29,582
1952	6,295	10,128	545	16,968
1953	7,300	11,430	555	19,298
1954	5,375	13,960	1,073	20,408
1955	12,530	12,868	1,112	26,510
1956	25,068	14,111	1,215	40,394
1957	34,561	13,193	1,333	49,087
1958	43,439	13,856	1,069	58,364
1959	50,605	15,493	1,517	67,615
1960	56,211	17,566	1,572	75,349
1961	58,840	17,800	1,589	78,229
1962	64,325	18,414	1,371	84,110
1963	56,477	16,695	1,254	74,426
1964	42,115	14,581	1,411	58,107
1965	77,705	17,674	1,381	96,760
1966	92,717	19,379	1,372	113,468
1967	111,718	19,905	1,551	133,174
1968	125,517	22,555	2,228	150,300

Source: Škoda-Auto Historical Archive (Aša), Mladá Boleslav, Automobilové závody, národní podnik (AZNP), b. 34 Český Státní statistický úřad, zprávy a rozbory, rok 1970, *Rozvoj automobilového průmyslu v ČSR*, tab. 13.

According to the president of the Czechoslovak Automobile Club:

> Driving is no longer a privilege of people who exploit workers, as in the times of the Republic. The data shows that now our cars go directly into the workers' hands [...]. The impetuous entry of the working class into motorization forces us to prepare our drivers for the various tasks connected to the defense of the nation.[20]

The pages of *Auto* and *Svět motorů* tried to justify the industrial policy choices of the First Five-Year Plan, which focused on the development of heavy industry at the expense of consumer goods, by assuring that the expansion of heavy industry would create the material and productive foundations for an improvement in the living standard of workers. This, in turn, would allow car ownership to be a commodity "delivered directly into the hands of the workers"[21] rather than a "symbol of class"[22] prerogative of "exploiters" for whom "normal people" had to step aside, as it had been in the past. Finally, if here and there these pages made veiled hints that the price to pay for the changeover to socialist production might be a certain delay in motorization, they were also confident that along with quantity of production, the world-renowned quality and perfection of Czechoslovak cars would be guaranteed by collaboration between technicians and workers.[23]

For this reason, ČZAL specialists stressed that while following the indications of the plan, it was still necessary to continue research and development for a new model with a more suitable selling price, maintenance, and running costs than the models currently in production. At the right moment, this new model would be able to respond to the population's insistent demand for a *malá lidová vozítka* [small people's car]. In any case, they were skeptical about the possibility to supply the "workers with a motor vehicle both useful and enjoyable", even over the middle term:

> At the moment, the domestic market has no importance, also because the composition of Czechoslovak society is changing. We speak in theory about the emergence of a new social class sufficiently well-off to be able to buy a car [...] and take care of its maintenance. We have to see when the right moment for this solution arrives. But as the production foreseen in both Five-Year Plans will not contribute either to a change in the number of vehicles in circulation or to their modernization, we prefer not to answer this question since [...] there will not be cars available for this new prospect.[24]

Discovering the Soviet Automobile Industry

The First Five-Year Plan allocated the automotive industry, and in particular automobile production, far fewer resources than Alexander Taub and the other technicians could have ever imagined. A study on the state of the transport industry by the National Statistical Institute in 1970 ascertained that there had been no progress in automobile production methods between 1948 and 1962 and that in those years, the main factory for the manufacture and assembly of automobiles in post-war Czechoslovakia, the Mladá Boleslav plant, had grown in a completely disorderly and random manner.[25] By 1949 the idea of building a new plant proposed by Taub in 1947 was totally implausible. Yet numerous notifications reached ČZAL from Mladá Boleslav concerning the need for an overall modernisation of its machinery, to be obtained by spreading investments over the five years of the plan or even by bringing in machinery from other plants. AZNP's management had made up a list of machinery considered by then unfit for continuous production, some of which had been purchased in 1928 when the *America* body plant was built, and had asked for its substitution with more modern machinery that existed in other factories but went unused.[26] Expenditure for these machines – principally Sunstrand and Le Blond lathes, Greenlee and ACME-Gridley automatic multiple spindle machines, Ingersoll cutting and drilling tools, Landis grinders, Heald automatic grinding machines, Gleason gear cutters, or comparable machines produced by Škoda Works – entailed a considerable financial outlay.[27]

At the end of 1948 there were still some who thought that 51 percent of these investments could be agreed and hoped that the "existing equipment"[28] could be supplemented with "British and Canadian loans, repairs, returns, with surpluses from domestic nationalized companies and by inter-company exchange of machinery".[29] In reality, after 1948 and until the late 1960s, the currency shortage and the western embargo that had meanwhile been imposed were determining factors that kept AZNP from obtaining the greater part of the requested machinery. In his memoires, Peter Hridlička, speaking of the early 1960s and the initial consequences of the embargo, recalls how in order to produce the new Gleason Format-type gears, he himself had to carry out all the technical measurements needed to use the old American machines imported before 1948 as it was impossible to buy new machines.[30]

A long document from 1950, "questionnaire for journeys to the Soviet Union", meant to be a sort of handbook for the six technicians who were authorized in 1951 to travel to the USSR and which consisted of over two hundred entries, gives an accurate description of the state of automobile production in Czechoslovakia.

It confirms that the manufacture of cars in the Mladá Boleslav plant entailed a considerable waste of time and materials, and that in some shops the obsolescence of the machinery made it impossible to produce efficiently or stay on track with the objectives set by the First Five-Year Plan.[31]

The technicians who wrote this document were highly competent and aware of the progress made overseas in the automobile industry. They were posing questions to the Soviets at the same time as they described the state of production in Czechoslovak factories in light of what they knew about state-of-the-art technology. The questions and answers were divided according to departments: the body and casting shop had the majority of pages, but the AZNP technicians also focused on the movement of materials between shops and the quality of materials used. The notes on wage systems and the organization of work were especially full of information, particularly as regards the presence and employment of female workers.[32]

The technicians started out from the diagnosis of the state of production that had been made by Taub and the Czechoslovak technicians who had gone to America.[33] The backwardness of body manufacture was considered to be the main problem of the Czechoslovak automobile industry. The body type was also outmoded compared to both American and European standards: neither the self-supporting chassis nor the all-steel body had as yet been introduced in AZNP. The Czechoslovaks were aware of this, but they also knew that the introduction of innovations of this kind would require a reorganization of the lines following principles of mass production, which would make it impossible to diversify body construction or might even necessitate the creation of a completely new body shop.[34]

The AZNP technicians also took care to examine the quality of human capital – how the Soviet technicians had been trained, whether in the field or in specialized schools, and what they based their updates on, technical literature, motion studies, or calculations. Unfortunately, in these cases only rarely did they take the trouble to describe what was happening in Czechoslovakia.

Other necessary information regarded the handling and storage of forged parts as well as their quality: at Mladá Boleslav, transport was done on trolleys and not by conveyor belts; as for the quality of the press work, workers had to make continual adjustments to straighten out the majority of pieces.[35] The result was that total working time was inordinately longer than it should have been. The small-parts presses were loaded by hand, with few exceptions. Waste material that fell on the floor was also recovered by hand: it was swept up and carried away on trolleys by unskilled workers. The workforce on small and medium-sized presses (10-250 t) was 50 percent female, while on large presses (300-1000 t) it was only 20 percent female. In AZNP there was no special protection against accidents except, in some cases, recourse to pincers on smaller and older presses and to a double start-up for large presses.[36]

Welding was done in 90 percent of cases with spot welders, following the method introduced in Europe in the 1920s.[37] But, according to the authors of this document, even this type of operation was difficult. In fact, after extracting the pieces forged by machine, additional time was needed to achieve precise dimensions: cleaning the welded spots with a steel brush took 15 minutes, adjusting the welds with an oil-soaked wood hammer took 150 minutes, and 40 minutes were needed to smooth the welded surface with a manual rasp and bathe it with antirust. Leftover materials fell on the floor, were picked up by a worker, and then melted down and reutilized. Finally, the adjustment of the bumpers and the hood required another 155 minutes.

In painting and finishing there was a great gap between an American shop and AZNP: at Mladá Boleslav, alkyd resin paints, lapping, and super-finishing, which represented the latest novelties in America, did not exist. The paint used was nitrocellulose, and smoothing was done by the abrasive stone method (honing). Along with the operation of polishing the body, this was entrusted to women, 50 percent of whom were between forty and sixty years old. They used air-blowing instruments with felt and lambskin discs to clean and degrease all surfaces.

In the paint shop the bodies slid along the floor on trolleys and not, as the technicians thought better, on ceiling conveyors. This meant that the bottom of the trolleys was dirtied by the paint that was gun-sprayed four times from above.[38] The paint was then dried in an electric chamber for fifteen to eighteen minutes at 1000°C and for two hours at 70°C in a steam drying chamber. To protect themselves from the toxic fumes of the paint, the workers at times wore masks; in some cases they used an oxygen tube. As with the press shop, the technicians understood the inadequacy of these measures and hoped for more modern solutions.

Another bottleneck that caused parts to stay in the body shop even longer lay in the press shops – in fact, there was no written material in Czech concerning large-sized dies. The arrival of the presses requested for 1949 had improved the situation; parts could now be forged with multiple operations performed mechanically, and not as before on a single machine followed by a worker's intervention. However, for copiers there was still the problem of shortage of machinery. The Pratt & Whitney "Keller" tracing machine with electromagnetic chuck partially solved Mladá Boleslav's problems.

In the machining shops the limits imposed by a shortage of machinery did not seem as insuperable as in the press shops, and the technicians were more optimistic.

There was a machine every nine square meters, the machines were bolted to the shop floor so that they would not be moved by vibrations, and belt cables for electricity (the machines did not have a single engine, they still depended on a central electric engine) ran along the ceiling. In channels behind the machines there were tubes for the distribution of cooling liquid. Sunlight shone directly into the shop because it had not been possible to isolate it properly; artificial lighting was supplied by central overhead lights and by low-tension lamps near the machines.

The arrangement of the machines reflected the coexistence of outmoded and modern methods: there were both machine tools of the same type grouped together and sequential lines for more complicated parts. There were semi-automatic machines as well as modular systems, but transport inside the shops and the utilization of warehouses suggested the existence of a considerable degree of approximation and making-do. This was also revealed in the pages dedicated to the control of materials and products.

Production was planned for two shifts, but because of the shortage of machines, repairs had to be done during working hours, creating slow-downs and production blockage. The workers carried in the tools for making adjustments after taking them from central warehouses and put them into a box beside the machines they operated.

Along with the shortage of machinery, the poor quality of the materials both for the parts to be manufactured and for tools was an additional problem. At Mladá Boleslav, gear wheel teeth were manufactured with sets of cutters produced in the U.S., and methods of shaving were gradually being introduced. The question of the material used was relevant because at AZNP they did not use nickel and molybdenum, as in America, but chromium steel from the Škoda plants in Plzeň and Poldi. This material would buckle even after planing and grinding, and so cause the gears to block continually. The same problem arose when the question of the durability of Czechoslovak-made utensils was examined: for example, Czechoslovak milling cutters and shaving utensils were subject to far more wear and tear than those produced abroad, and for this reason they had to be sharpened and replaced more frequently. [39]

Nor was AZNP advanced in car assembly. Changes in production plans occurred frequently, as did interventions regarding the workforce. All in all, the lines (on which different models were still assembled on the same day) needed to be speeded up and mechanized.[40] From the point of view of manufacturing technology, Mladá Boleslav was a modern 1930s-style shop. The machinery was old, but not obsolete; recourse to manual labor was too high, over 60 percent against 40 percent working on machines. They tried to introduce new organizational methods in the production process and new types of manufacturing, but there were bottlenecks due to machinery shortages, the poor quality of materials, and the constant shortage of workers.

The document prepared for the Soviet missions leaves few doubts about how obsolete the Mladá Boleslav factory was gradually becoming, but at the same time it shows that local technicians possessed impressive know-how and a clear ability to pinpoint problems and indicate solutions that could increase efficiency when faced with shortcomings in the production process.[41] In 1950 the Czechoslovak technicians still felt that the plans drawn up with Taub in 1946-48 represented a starting-point for developing the sector, and they considered the American model the main standard of efficiency for evaluating the steps to take in Czechoslovakia. Indeed, the wording of the document suggests that, besides wanting to show their know-how to the Soviets, the authors considered what they had learned in America to be the reference point for evaluating the supposed modernity of the Soviet industry. In this context, much of the documentation produced in these and subsequent years reveals the Czechoslovak technicians' insistence on the peculiarities of the history of their automobile industry, which had until then followed a path of development not very far from Europe and America, but profoundly different from the Soviet Union. As they wrote, the Soviet industry had not developed through a process of "natural selection"[42] leading to concentration in the sector, as had happened in Europe and America; rather, it had been created by the state in relatively recent times, and its production had always been organized and regulated by a plan. Consequently, the features that marked the development of the Czechoslovak automobile industry and the know-how at its disposal made it seem natural to turn to the experience of western capitalism and to what the technicians had observed during their journeys, when the factory doors were still open.

Although political and diplomatic agreements with the Soviet Union were increasingly binding, Czechoslovak technicians did not have the chance to visit Soviet auto plants until 1951. In the first months of 1949, negotiations on a possible cooperation between the Soviet and Czechoslovak automobile industries were still going on, regarding the number of Czechoslovak technicians who would be allowed access to Soviet factories, what they would be shown, and above all under what conditions.[43] In fact, the first visits to the USSR by Czechoslovak automotive industry technicians did not take place until February to April 1951, when six ČZAL engineers and the directors of some major production units visited the Ministry of Automobile Production and Agricultural Equipment, the Stalin automobile factory in Moscow (ZIS), the Molotov plant in Gorky (named Molotov from 1935 to 1956 but better known as GAZ, Gor'kii Automobile Factory), the ATE-1 electrical equipment plant in Moscow, and the Technical Museum in Moscow.[44] Among the members of this delegation was Stanislav Císař, a faithful Party member who had been appointed as director of AZNP in 1950, but who had no particular technical qualifications on record. It should therefore be stressed that in the years before 1951, and probably afterwards as well, Czechoslovak technicians' knowledge of the organization of production in Soviet factories was very sketchy and based chiefly on propaganda. According to a note dated March 9, 1949, Soviet specialists had already visited Czechoslovak automotive plants in 1947; in 1948 Czechoslovak technicians had to answer detailed questions posed by the Soviets on the state of their production. Nevertheless, what Czechoslovak technicians knew about the Soviet motor vehicle industry came from Soviet and American magazines and, later on, from trade fairs and shows.[45]

Unacquainted as they were with the real state of the Soviet automobile industry and having as yet only a vague understanding of the goals and methods of initial Sovietization, in 1949 Czechoslovak specialists thought they could use organizational methods and practices deriving from their own experience and their trips to America to carry out the new objectives of the Five-Year Plan. But the importance they assigned to the rationalization of production, which they saw as a way to consolidate and enhance the value of the national automotive industry, conflicted with political intentions, partly dictated by Moscow, to downsize the country's automobile production. It is not clear to what degree the specialists were aware of the growing gap between their ideas and projects of technical and industrial modernization and the outcomes of the measures being taken in the "upper spheres" of the planning apparatus. It is clear, however, that their plans for the development of the sector, and above all the search for the more efficient organization that was key to this development, came into conflict both with local resistance to the centralization of the decision-making process that was the Party's goal and with Moscow's still confused ideas about the integration of Comecon and the division of production among the People's Democracies.

Mass Production and Comecon Challenges

Temporarily resigned to the impossibility of obtaining the necessary machinery, the ČZAL technicians thought they could achieve some of the advantages of mass production by adopting certain organizational measures which had been formulated by Taub and did not seem to contradict the directives of the Plan. This was how they intended to supervise the processes of sectorial concentration, product rationalization, and standardization and growth in labor productivity. What was involved, they felt, was finding the best way to facilitate and accompany the process of "natural selection" that had elsewhere allowed companies with greater "financial possibilities and technical experience" to prevail over others on the market.[46]

In what was clearly taken from Taub's reports, the technicians envisioned a single, centrally managed combine made up of the country's most modern and efficient plants. Their reference points were the American and British experiences: the American market had been dominated by large-scale corporations, while in the U.K. after the war, the sixty-three pre-war manufacturers formed six consortia in which 90 percent of production was concentrated in order to equip themselves with modern machinery and reduce selling prices. The results were judged "sensational".[47]

From this standpoint, the Czechoslovak industry, being nationalized and centrally managed, had all that was required so that its structure and the size of its production units could fit the needs of more efficient and less costly production.[48] It was envisioned that Czechoslovakia would reach the end of the First Five-Year Plan with a single program to develop the national automotive industry. This would allow for the manufacture of a range of models using common, interchangeable components and assembly groups that would be large enough to satisfy the different kinds of demand – including defense and foreign trade – from the standpoint of price, power, and load. In this case, the model was the organization of military production and the war industry. Rationalization was also to involve the manufacture of components: each of them was to be standardized and centralized in a single, specialized factory. As the American experience of General Motors showed, only in this way would it be possible to achieve high volumes of production and modern methods.[49]

ČZAL intended to eliminate the semi-autarkic structure of the individual factories, which were still manufacturing almost every part in-house; it proposed to identify the most modern plant for each component and to concentrate large-scale production there. The Czechoslovak automotive industry should function at the organizational level like a great American corporation, or better like the British Big Six, whose strength lay in the single private company's readiness to delegate power to the center.[50]

In reality, ČZAL's rationalizing project came up against the resistance of those who were responsible for the single production units and of local Party functionaries. Four years had passed since the factories had been nationalized and placed under the authority of ČZKS-ČZAL, the old industrial complexes broken up, and their workforce transferred; yet it was still the "local patriotism" and the "desire for isolation" of the former private enterprises and producers that constituted the most important obstacle to the rationalization of the sector.[51] This was, paradoxically, the factor that most strongly distinguished the nationalized automotive industry from British private enterprises, which on the contrary accepted the use of common offices and standardized parts and also reached an agreement on commercial policy:

> we face significant obstacles from the local patriotism of individual companies, [...] their desire to remain isolated is far greater than what we find in private foreign companies which, for example in England, have agreed to collaborate even though they are not united by the constraint of nationalization.[52]

What was happening in Czechoslovakia was something very different from the description by the technician who had written this document. The relations concerning the delegation of authority that underlay agreements among the British Big Six or between the central management and the single divisions of an American corporation were undoubtedly complex; but what this technician was reporting was the far greater problem of asymmetry of information, the deformation of the flow of information from bottom to top, which would continue to plague the command economy, and which in 1949 was already creating difficulties for Czechoslovak planning.[53] In any case, the solution proposed by the author was far different from the well-known system that would in fact be used by planners to overcome local resistance, a system that proved to be as ineffective as it was repressive.

According to ČZAL technicians, the communication problems between production units and the "upper sphere" offices could be solved by creating a central development department. This would coordinate new construction and the standardization and unification of shared parts, guaranteeing that they really were utilized in all plants, and at the same time would interact with the plants to formulate ideas for future development. This center, which would require trained personnel and independent means, would aim at doing scientific research as well as planning. It could also perform the fundamental activity of documentation, cataloguing magazines and collecting "technical indications, prospectuses, notes and other papers offering information on the technical features, defects and market position of a particular model",[54] which would then always be available if needed during meetings or negotiations.

Through this "development department", ČZAL would take over the management of the main assembly plants as well as auxiliary plants and component parts manufacturers. It would also supervise research and patent activity and would have a statistics office and a central documentation center that could gather information on the worldwide transport industry. This was an ambitious project, considering the model which the development sector seemed to refer to, i.e. the research and development departments of the American "big corporation".[55]

But when a development department was in fact founded in 1952, it took on features far more like the Soviet NAMI (Scientific Automotive Institute) than development departments in Vauxhall or Chrysler. Indeed, the *Ústav pro výzkum motorových vozidel* ÚVMV [Motor Vehicles Research Institute], as this center was called, had no real decision-making functions or power regarding production or coordination.

It should, however, be noted that, despite a few brief references to the Soviet experience and the use of language conforming to the *tòpoi* of "building socialism", the guidelines for developing the sector were still derived from the American model and Alex Taub's consultancy. In reality, these technicians showed no awareness of the fact that at a technical level, the Soviet model and the American model might represent incompatible alternatives.

Corresponding to the projects for rationalizing the national automobile sector, there also was an interesting project for international product specialization. This project represented the response of ČZAL technicians and the Czechoslovak government to Soviet attempts to limit or even cancel automobile production in the country.[56] ČZAL's project can be seen as an example of the difference between the muddled Soviet intentions and the understanding of the technical consultants of the Czechoslovak Party's leadership on how to go about effecting economic integration in the Soviet bloc. At the January 1949 conference in Moscow that founded the Council for Mutual Economic Assistance (Comecon), the representatives of the Czechoslovak government strongly upheld the need to make Comecon an international organization whose passwords would be integration and specialization, and which would eliminate industrial autarky and parallel investments. They envisioned a division of tasks whereby Czechoslovakia was to specialize in heavy engineering and, as regards the motor-vehicle industry, in the production of trucks from seven tons upwards.[57]

Czechoslovakia was the most industrialized member of Comecon, and it was also the country with the most modern engineering industry.[58] It was also highly dependent on the import of raw materials and metals, especially iron products and aluminum, as well as certain types of machinery from the West.[59] In 1949, the possibility to continue to receive these supplies, which were vital for the economy of the people's democracies, was jeopardized by the embargo imposed in 1948 and above all by the compilation of the list of the Coordinating Committee for Multilateral Export Controls (CoCom) in November 1949.[60] What is more, the lack of strong currency, a chronic problem for the Czechoslovak government, made it difficult to purchase supplies even on the friendly markets of neutral countries.[61] For this reason, in Czechoslovakia the possibility of intraregional cooperation underlying the creation of Comecon was welcomed with particular interest: the hope was that in this way it would be possible to access raw materials and Soviet technology, which at the time seemed a more reliable alternative than western sources. The Czechoslovak government was therefore deeply committed to promoting the intraregional cooperation that Comecon seemed to promise; in particular, it fought to achieve genuine economic integration, which it thought would combine the rerouting of trade with a real international division of labor and production programs.[62] Because of its still relatively close economic relations with the West, Czechoslovakia was also subjected to considerable pressure from Moscow, which expected an immediate cessation of trade relations between its allies and capitalist countries.

The first meetings of Comecon in Moscow and Sofia (April-August 1949) took place in a climate of deep mistrust and even real fear on the part of members. In his letters to Prague, economist Josef Goldmann (1912-84), the Czechoslovak representative, reported not only a lack of even the vaguest idea of the state of industry in other member countries, which made it extremely difficult to develop common projects, but also the impossibility for him to leave the Soviet Union or to communicate the contents of the discussion to Prague. Between February and April 1949 there was an intense exchange of letters between Goldmann, Ludvík Frejka (1904-52), economist and head of the presidential offices, and the organizer of the Taub affair, František Fabinger concerning the coordination of production and the exchange of technical and scientific knowledge among Comecon members.[63] This documentation confirms the growing uneasiness of the Prague authorities caused by the effects of the embargo and the hopes that Goldmann and his colleagues placed in the Comecon project of integration, which they saw as the ideal solution to the closing off of the West.[64] However, it was not at all clear what Moscow was aiming at; what is more, the member states held very different positions, which reflected the dissimilar economic conditions in their countries.[65]

In a situation where everyone was competing against everyone else, with each member state trying more or less covertly to gain Moscow's favor, Goldmann and Fabinger's goal was to achieve efficient coordination of planning on an international scale that would lead to the international division of work – of production programs – and eliminate parallel investments in the people's democracies, which they considered to be a waste of time and resources. In this way, the Czechoslovak industry would obtain the resources necessary to produce efficiently for the entire Comecon block, at least in the key sectors of chemicals and engineering. Goldmann was also aware, however, that if this did not happen, the general appeal for technical and scientific collaboration as a first move towards economic integration would turn into a flow of patents and know-how from Czechoslovakia to the less-developed countries. Then Czechoslovakia would very soon lose its industrial leadership and have to face dangerous competition within the eastern bloc itself.[66] Moreover, the industrial development of Bulgaria and Rumania would make the raw materials Czechoslovakia needed for its manufacturing even scarcer.[67] It was also clear that if Czechoslovak objectives were not met, the country would have to sacrifice some of its most advanced product lines on the altar of socialism, which were also the ones the people cared about most, including the automobile.

The planners' defense of the automobile also had its economic rationale: in the years after the war, over 75 percent of auto production was exported, and therefore it represented a source, albeit a limited one, of the strong currency needed to reach the Plan's production objectives for heavy and strategic industry. For this reason, too, the division of production programs in the automotive sector, above all the automobile, represented an important element in the match that the top Czechoslovak planners played and lost in Moscow. Between September and December, ČZAL busied itself with the formulation of a project that could represent the basis for a "politics of the automobile" in Comecon, and that would justify the survival and expansion of Czechoslovak product lines.[68] According to ČZAL, it was likely that the number of commercial vehicles and automobiles in the territory of all the states of the eastern bloc, with the exception of the USSR (Hungary, Rumania, Poland, Bulgaria, and Yugoslavia), was lower than it had been in post-war Czechoslovakia, and that the products were far inferior in quality. But in August 1949 the exact figures measuring this gap were not known. The obstacles in the way of a real apportioning of production programs were many. First of all, it was necessary to obtain precise data relative to the domestic market and the productive capacity of each single country. Then, the Czechoslovak technicians also realized that in some of the people's democracies, there was a strong tendency to develop an autonomous automotive industry or even to buy foreign licenses from western countries. The Czechoslovaks were informed, in fact, about the negotiations between Fiat and Polish authorities.

ČZAL technicians felt that to build plants in every country was useless if not harmful, because "the real sales data in the states of the eastern bloc" taken all together did not seem to offer adequate market prospects for even one modern automobile company.[69] There were also strategic political reasons that made this an unwise idea: the army had to be able to count on high-volume, standardized vehicles that could be assembled in every country.[70]

Product specialization could even be extended to components. The idea of a single, hyper-specialized plant for each component was extended to the whole eastern bloc, and "a mosaic production" for the entire area of the Comecon was envisioned, with each country specializing in a specific part or model.[71] The Czechoslovak specialists were convinced that their own automobile industry, alongside meeting its own needs, could easily cover the real demand from the entire Eastern bloc countries with its existing factories, and maintain 50 percent export.[72]

Being aware of Czechoslovak superiority in the production of motor vehicles, ČZAL offered to play the role of guide and in the meantime developed a strategy for the modernization of its own facilities. To produce a "people's car" for the whole eastern bloc meant having a market for volumes of production that could justify the increase in investments and the purchase of machinery that would be indispensible for mass production.

Moving from a capacity of 33,000 vehicles a year, which had been reached 'only' thanks to the expansion of the second shift and was still not adequate to meet national needs, to 50,000 vehicles a year, which was advisable to lower costs further, could only be achieved through larger investments and the introduction of special, single-purpose machines.[73]

These same observations were made even more pointedly concerning the question of how to divide production programs for commercial vehicles among Comecon countries, especially for the seven- or ten-ton trucks. These trucks were extremely costly to produce because of the low quality of the materials employed and because they were manufactured in small series, a method suitable for batch production that required more workers and higher pay. Abroad, automated lines and transfer machines were used for the same models, which led to sizable cost advantages.[74]

Finally, the document mentioned a more comprehensive plan formulated by ČZAL together with the Minister of Defense, which envisioned the production of twenty-five different models of military-commercial vehicles, twelve for civilian use. They would be built using only fifteen types of engines, four gear-changes and seven gear boxes, and would be produced in the various people's democracies. By taking advantage of the enormous potential of Comecon, it was possible to realize economies of scale and diversification over a huge area. But as the head of planning objected to the managing director of ČZAL, the political problems that had to be faced were as huge as the area involved.

Table 3.3 – Automobile Production (in thousands of units) per Year in the Eastern Bloc, 1937-68 (including ambulances and special vehicles)

Year	USSR	GDR	Poland	Hungary	Rumania	Yugo-slavia	Czecho-slovakia
1937-38	5.5	60.8	1.9				13.3
1950	64.6	6.2					24.4
1955	107.8	22.2	4			0.8	12.5
1956	97.8	28.1	5.8			1	25
1957	113.6	35.6	8			3.1	34.5
1958	122.2	38.4	11.5			2.9	43.4
1959	124.5	52.7	14.2			4.4	50.6
1960	138.8	64.1	12.9			10.5	56.2
1961	148.9	69.6	14.4			15	58.8
1962	165.9	72.2	16.1			13.1	64.3
1963	173.1	84.3	18.3			20.9	56.4
1964	185.2	94.1	20.6			27.2	42.1
1965	201.2	102.9	26.4			35.9	77.7
1966	230.2	106.5	29.2			33.7	92.71
1967	251.4	111.5	27.7			42.3	111.7
1968	280	115	40.4			60	125.5

Source: Škoda-Auto Historical Archive (Aša), Mladá Boleslav, Automobilové závody, národní podnik (AZNP), b. 34. Český Státní statistický úřad, zprávy a rozbory, rok 1970, *Rozvoj automobilového průmyslu v ČSR,* tab. 34.

Table 3.4 – Automobile Production (in thousands of units) per Year in the Eastern Bloc, 1937-68

Year	USSR	GDR	Poland	Hungary	Rumania	Yugo-slavia	Czecho-slovakia
1937-38	141.5	79.9	3.7				16.1
1950	362.9	8.2	0.8	2.9		0.8	31.4
1955	445.3	34.1	14.8	5	3	3.2	26.5
1956	464.6	45.3	19.1	3.7	5	4.2	40.3
1957	495.4	51.1	23	4	5.1	7.2	49
1958	511.1	54.5	25.3	5.2	6	7.7	58.3
1959	495	67.4	32.2	4.9	8	8.6	67.6
1960	523.6	76.4	34.4	4.5	10.4	16	75.3
1961	553.3	81	26	5	12.4	21.2	78.2
1962	577.5	79.7	41.1	5.8	12	24.4	84.1
1963	587	92.9	48.4	5.4	17.4	29.9	74.4
1964	600.6	103.3	52.8	5.1	19.4	38.7	58.1
1965	616.3	117.5	56.6	5.5	20	45.5	96.7
1966	675.3	126.6	66.9	8	19	46.4	113.4
1967	728.8	133.4	68.3	8.3	20	55.6	133.1
1968	801	138	84.9	8.5	20	74.9	150.3

Source: Škoda-Auto Historical Archive (Aša), Mladá Boleslav, Automobilové závody, národní podnik (AZNP), b. 34. ČeskýStátní statistický úřad, zprávy a rozbory, rok 1970, *Rozvoj automobilového průmyslu v ČSR,* tab. 34.

Soviet Advisors and Tatra Plan

In September 1949 the Czechoslovak government "invited" in the Soviet advisors.[75] They would preside over the transfer of Soviet technical and organizational know-how to Czechoslovakia and its application, as well as transform what had been a project for cooperation into what immediately emerged as a unilateral project for the "spread of Soviet science".[76] The advisors arrived in the summer of 1951. Their role was half-way between that of business consultants (well-paid and reimbursed by the Czechoslovak government) and military experts who answered only to Moscow. Their intervention did not concern technical questions or production programs, as their hosts had expected. Rather, its goals were far more ambitious: to eliminate the last residues of capitalism from the economic apparatus.

The Soviets organized the reform of the SÚP [State Planning Office], the transformation of central directorates into central administrations inside the ministries, and the expansion of the *Ministerstvo všeobecného strojírenství,* MVS [Ministry of General Engineering]. They also convinced local planners that the Czechoslovak economy possessed material and human resources that were not adequately exploited. Thus, in every sector the imperatives were to eliminate new investments; to utilize existing reserves by applying the Soviet rule of maximum plans; and above all to impose discipline, cooperation, and order inside factories and in relations among the various production units. In the end, for the automotive industry, the reform of planning and industrial management translated into the abolition of ČZAL, the total militarization of the sector, and the collapse of Czechoslovak automobile production.[77]

This perhaps gives an idea of the profound misunderstanding of the Soviet intentions nurtured by the Czechoslovak technicians and politicians, who had looked to the Soviet Union with great expectations. On the other hand, it is clear that the particular post-war political and intellectual climate greatly contributed to the blindness of the technicians and planners: they believed that the Soviet model of industrial organization would guarantee efficiency and development, which would in turn guarantee the survival of the nation and the creation of Czechoslovakia as a great industrial power, albeit under Stalin's aegis. In reality, in the automobile sector (among others), things were to go very differently indeed.

The epilogue was dramatic: as Karel Kaplan has written, the farewell reception for the Soviet advisors in November 1951 ended with the arrest of Rudolf Slánský, who had presided as vice Prime Minister responsible for the economy.[78] The numerous trials that followed the death penalty he received involved many of the men who had been responsible for planning in the years 1948-51. Among them were Josef Goldmann, Ludvík Frejka, and František Fabinger, who were all accused of slowing down the transformation of the Czechoslovak economy into a socialist economy and of not cutting off relations with the West.[79]

Although ČZAL's proposal and the idea of specialized production advocated by Czechoslovakia did not win Moscow's favor, in April 1950 there was a Soviet intervention in the Comecon bureau regarding the production programs of the Czechoslovak auto industry.[80] On this occasion, the Soviets suggested concentrating in Czechoslovakia the production of heavy trucks (from seven to ten tons) destined for the entire Comecon block and the production of only one automobile model in Czechoslovakia, the Tatra 600.[81]

Various objections to this proposal were raised by ČZAL technicians. Partly in response to expectations originating in the factories, they favored the idea of producing two automobile models in Czechoslovakia: the Tatra 600, which was particularly suitable for official representation and would substitute the Tatra 87; and the Škoda 1102, which in the future would be substituted by a new prototype, the Š1200, a car that could satisfy standard needs and be assembled in various Comecon countries.[82] In this framework, Czechoslovak production would integrate the production envisioned for Poland, a 1,300-cylinder auto on Fiat license, and in Hungary, where a small vehicle was planned.[83]

Nonetheless, in November 1950, the SÚP announced the launch of the Tatra plan. Czechoslovakia would produce a large-size truck, seven to ten tons, and the Tatra 600, which has been produced in Kopřivnice since 1948.[84] Only a few days later, another order arrived from the Economic Council of the Central Committee of the Czechoslovak Communist Party:[85] to transfer manufacture of the Tatra 805, a light off-road military vehicle, and the Tatra 600 to the AZNP in Mladá Boleslav and at the same time to phase out the Š1102. This seemed to mean postponing manufacture of the Š1200 indefinitely.[86] The Tatra plants in Kopřivnice would be totally dedicated until 1956 to the manufacture of trucks and the heavy off-road military vehicle, T128. Following the letter of December 8, 1950, the Minister of Industry ordered ČZAL's general management to modify the production program plants and to communicate the new plan for the year 1951-52 to those responsible in AZNP.[87]

Thus, the project to rationalize and concentrate the manufacture of a single automobile model in a single plant on national territory was finally carried out, but it happened on a basis far different from what the ČZAL had envisioned. Military requirements, which had made it necessary to set aside the greater part of the productive capacity of auto and mechanical plants to wartime production, continued to impose serious constraints on the production of automobiles. The plan on which the changes in production programs and the division of labor were based was taken from an accurate study of how the Germans had organized the factory and the workforce and how they had exploited AZNP's potential during the occupation; it also foresaw further reductions in civil production.[88]

Nevertheless, the Tatra plan remained an ambitious project: for 1951 it envisioned the manufacture of 16,600 Š1102 vehicles and 2,000 T600s; in 1952 it would manufacture 12,000 vehicles. However, these targets were impossible to achieve, and by the end of 1953 nothing remained of the prospective plan for the years 1953-55, which had envisioned a total production of 38,000 T600s. To make things worse, the costs of the operation were huge.[89] The result was that in 1952-53, Czechoslovak automobile production plummeted to minimum levels never before reached: a mere 6,300 vehicles a year, with only 2,100 Tatra 600s produced in Mladá Boleslav.[90]

This crisis and the impossibility of implementing the plans had been foreseen and reported by Mladá Boleslav's experts and workers even before the SÚP's approval of the transfer of the Tatra plan from Kopřivnice to Mladá Boleslav. Indeed, as early as November 1950, AZNP had presented the first notes criticizing the project to the Ministries of Industry and Defense and to the SÚP. The technicians' perplexities concerned various aspects of this move, but what most worried them was the arbitrary decision to substitute the Š1200 with the model T600. They evidenced both the difficulty in finding the necessary materials and in setting up production of the new vehicle: they stressed that compared to the Š1200, the T600 required a significantly greater quantity of light metals and alloys which, in their opinion, should have been used for other important tasks for the state. They were convinced that the T600 was still not ready for mass production.[91]

AZNP's protests reached the President, Antonín Zápotocký, when his secretary forwarded to the management of ČZAL, which was at the time being disbanded, a letter received on May 31, 1951, from an unnamed "comrade", who was an employee of AZNP.[92]

In the letter, the "comrade" who had just visited Kopřivnice – where Tatras had been produced until then – stressed the irrationality of the decision of producing Tatras at Boleslav instead of Škodas 1200. In his opinion, the 1200 was more or less equivalent to the old Tudor (Š1101-1102) in weight, price, and also the number of hours of work needed to construct a single car. On the other hand, the 1200 was an improved Tudor and could be included in the category of middle-sized engines without increasing production costs. There was also a problem of production times and costs of both production and maintenance:

Table 3.5 – Automobile Production (in thousands of units) in Selected Countries Classified by Engine Power, 1960-68

Country	1960	1962	1964	1966	1967	1968
FRG						
Up to 500 cc	31	22	15	12	9	7
501-1,000 cc	265	252	439	73	86	87
1,001-1,500 cc	1136	1363	1611	1584	1210	1554
1,501-2,000 cc	285	382	481	722	572	653
More than 2,000 cc	100	90	104	124	166	234
Total	1817	2109	2650	2515	2043	2535
Great Britain						
Up to 1,000 cc	1058	403	391	301	289	258
1,001-1,600 cc	235	263	267	413	319	363
More than 2,800 cc	60	51	64	76	45	61
Total	1353	1249	1868	1604	1552	1816
France						
Up to 1,000 cc	678	918	729	893	845	
1,001-1,500 cc	351	186	364	610	642	
More than 1,500 cc	146	236	257	283	290	
Total	1175	1340	1350	1786	1776	
Italy						
Up to 500 cc	126	177	249	324	399	
501-1,000 cc	245	304	389	471	440	
1,001-1,500 cc	188	343	324	441	476	
1,501-2,000 cc	17	30	51	38	118	
More than 2,000 cc	20	24	16	8	6	
Total	596	878	1029	1282	1439	
Czechoslovakia						
Up to 988-1,221 cc	55	63	41.6	92	119.9	124.3
More than 2,000 cc	1.2	1.3	0.5	0.7	0.8	1.2
Total	56.2	64.3	42.1	92.7	111.7	125.5

Source: Škoda-Auto Historical Archive (Aša), Mladá Boleslav, Automobilové závody, národní podnik (AZNP), b. 34, Český Státní statistický úřad, zprávy a rozbory, rok 1970, *Rozvoj automobilového průmyslu v ČSR*, tab. 37.

the time to produce a Tudor is roughly 240 hours, for the Tatra 700 hours. The 1200's metallic body can be mass produced, while this still has to be perfected for the Tatra. We also need to compare the costs and running expenses: the Tatra will cost more due to its depreciation and greater fuel consumption. Today's customers have to have their say and so we have to consider the heavy consumption of light metals, 68 kilos for every car, while the 1200 allows the use of only grey iron and some molds without compromising quality. Heating is still unsatisfactory in the Tatra – cold

> in the front in winter as the engine is in the rear and air-cooled, which often causes wear and tear on the cylinders. We could say more, but this is enough for now.[93]

The clarifications requested by Zápotocký's spokesman a few days later, on June 18, 1951, were provided by the General Director of ČZVV (formerly ČZAL), who guaranteed that the situation was completely under control and that the colleague from AZNP, motivated by his intention to support as best he could the position of the Mladá Boleslav AZNP, had undoubtedly exaggerated the significance of his statements. And in any case, he added that the decision was actually made "by the Presidency of the Economic Council in the meeting held on June 6 of this year, with the best of intentions, corresponding to the requests and opinions not only of SÚP and ČZVV but also of Tatra and AZNP workers".[94]

The worst predictions came true during the transfer of production from one factory to the other. On January 5, 1951, a temporary contract was agreed at Kopřivnice in the presence of representatives of both companies, which bound both parties to accept the transfer. The final contract was only drawn up some months later, on August 23, 1951.[95]

Meanwhile, all the problems that were to convince the director of Mladá Boleslav to ask for a change in production programs gradually emerged: Tatra could not ship the majority of the machine tools needed to prepare the manufacture of the T600 to the AZNP because they had been used to manufacture the T128, and the same was true for the steel alloy materials required for military production. What is more, the difference in the manufacturing process used in the Mladá Boleslav plants, continuous even if not highly mechanized, and that used at Kopřivnice, by batches and crafts-based, made it necessary to modify most of the production processes at Mladá Boleslav. This meant that it was extremely complicated to produce the T600 with the machines available there. Adapting or adjusting tools was not enough – new ones had to be made. But the tool-shop at Škoda had been entirely turned over to military production.

The factory was up a blind alley. In a short time the dramatic situation regarding materials (light metals, especially class 320 silumin and steel alloys) and the need to exploit at least partially the productive capacity of the plants, which were slowly grinding to a halt, made it necessary to effect another change in production programs: in 1952 manufacture of the T600 was replaced with the Š1200, which remained the only model of automobile produced at Mladá Boleslav until 1956 and, with very few changes (Š440), until 1964.

The reasons for the transfer of the Tatra plan are still obscure (if we exclude the intervention of Moscow): there is only one document in the Škoda archives and ČZAL papers that allows a reconstruction of what happened to production at Mladá Boleslav in the years from 1951 to 1953, but it is dated 1954. Even if its contents are partially confirmed by the monthly minutes of the meetings of the managers of AZNP factories, above all as regards the material difficulties that caused the production blockage, it creates problems of interpretation, because August 1954 marked the beginning of the trial against those responsible for Czechoslovak planning, charged with and often convicted of sabotage and high treason. The names of Josef Goldmann and Ludvík Frejka appear in the documents among the names of those who did not support AZNP demands, thereby indirectly causing the failure of the plans.

This 1954 unsigned document alternates the desire to defend AZNP from the accusations of "local patriotism" and failure to respect the established goals with an unconvincing attempt to blame the "upper spheres" and especially those who seemed already condemned.

Chapter 4 Soviet Administration in Czechoslovak Factories, 1951-54

The Origins of Soviet-Style Labor Management

Institutional transformation and the intervention of Soviet advisors led to a semi-military economy. Instead of encouraging the use of modern machinery or innovative equipment, the authorities focused on making full use of the existing capacity and raising labor productivity.

From the point of view of economic policy, the Czechoslovak automobile industry was heavily penalized from 1949 to 1953, and production virtually came to a halt. The allocation of resources made it clear that the authorities had little consideration for this industry; at the same time, a violent attack was launched against worker autonomy and the work methods adopted in the Mladá Boleslav plant, an attack carried out in the name of organizational rationalization and increasing workers' productivity. This intervention, which peaked during the years of monetary reform (May 1953) and the definition of criminal regulations against absenteeism and staff turnover (June-July 1953), clearly served neither the demands of the market nor those of technological innovation. Instead, it did serve the new leadership's need for control and its policy of sovietization of Soviet Russia's satellite states.[1] Indeed, in these years certain Soviet managerial practices in the Czechoslovak factories resembled outright militarization of the workforce and can perhaps be seen as a step towards the affirmation of the Party's hierarchical control over the shops.

However, this attempt at totalitarian-style control encountered opposition and resistance of various kinds in Czechoslovakia. It was at least partly interrupted after Stalin's death and the German and Polish strikes in 1953-54.[2] Although in all likelihood politics, and especially international politics, had the greatest influence on the Czechoslovak "new course", it was, however, the near-sighted soviet-style industrial administration and organization of work that put considerable strain on relations between the Party and the class it claimed to represent. The 1953 strikes at strategic plants in Plzeň (Škodovy Závody) and Prague (ČKD), where the nation's most highly qualified workforce was concentrated, demonstrated an unmistakable sign of this strain.[3] There are no reports of major strikes at Mladá Boleslav. Still, analyzing the impact of Soviet labor management methods at AZNP's factory, the main automobile manufacturer in Czechoslovakia, lets us see just how alien Soviet principles were to the theories of labor management formulated by the Czechoslovak Taylorists in the interwar years.

Having gone through the 'natural selection' of a market economy, the history of the Czechoslovak automobile industry was profoundly different from the Soviet's, which was created and grew thanks to state initiatives in a planned economy. Furthermore, despite the common matrix in American scientific management, in the 1920s and 1930s the reworking of American organizational principles and practices had taken a completely different direction in Czechoslovakia from the one taken in the Soviet Union. As shown in the first chapter, in the interwar years the Masaryk Academy of Work (MAP) had made scientific management a topic of discussion among the Czech technical and business elites. However, MAP had not adopted a literal interpretation of Taylorism; instead, it valued the theories involving the "human factor" more and had translated Gilbreth, Emerson, and later on Mayo.[4] Paradoxically, while techniques of scientific management were being adopted in Europe by authoritarian political regimes in Nazi Germany and fascist Italy, where conflict was prohibited by the political context and factories were organized to shore up "authority", MAP was taking a stance very like those formerly held by the German Social Democrats and the "school of human relations".[5]

The militarization of the factories under German occupation had imposed a totally authoritarian version of Taylorism for the first time in Czechoslovak shops. Productivity did increase, but at the end of the conflict, these techniques were associated with foreign domination. Moreover, after the war the workers showed a voluntary commitment to reconstruction, motivated by their enthusiasm for newly won national liberty, along with the role played by factory councils in managing the post-war chaos, a role that proved decisive in laying the foundations for the consolidation of pre-Taylorist labor practices.[6]

Against this backdrop, "Fordist" infatuation and Alexander Taub's project for the reconstruction of the automobile sector emerged. The visits to America had strengthened the conviction of technicians and managerial personnel that American technology and productivity were politically neutral and wholly compatible with the Czechoslovak road to socialism contained in the Košice Treaty. As regards management systems, the model to follow was the work of the "Czechoslovak Ford" Tomáš Bat´a, who upheld the need to safeguard the workers' professional skills and integrity.[7]

Rationalization was, therefore, a complex issue which had been debated at length in Czechoslovakia under different political systems and which could no longer be reduced to a mere question of discipline after the horrors of the occupation. The admiration of workers and managerial cadres for Soviet successes coexisted with both a felt need for modernization and pride in the factory's history.[8] Despite the subordination of factory councils to trade unions that had been transformed into mere "transmission belts between the masses and the Party" and despite the anti-saboteur trials of 1952-53, in Czechoslovakia, unlike the Soviet Union, the pre-revolutionary past was not scientifically cancelled by totalitarian power.[9] Still, between 1950 and 1953, the campaign passwords for greater efficiency and productivity in socialist factories were discipline, piecework, and socialist emulation, elements fundamental to the Soviet version of scientific management, which in the early 1930s had diverged from the European tradition to take on forms suitable for an economy in a permanent state of war.

According to American writer Judith Merkle's analysis, the history of Soviet Taylorism can be traced back to the attempt to establish a revolutionary proletarian state, following Marxist methods: the proletariat was to use the state machine to create a wholly rational model of egalitarian distribution of industrial production. The problem was that the revolution broke out in a country where the industrial premises for socialist redistribution did not exist. Nor had Marx predicted the failure of the revolution in the rest of Europe and the establishment of Stalin's "socialism in one country". Consequently, the problem of building socialism became complicated indeed. In their search for technical and organizational solutions, the Bolsheviks resorted to a system originally conceived as a "philosophy of private business", the Taylorist system.[10] Indeed, Lenin himself had legitimized Taylorism in several of his writings, and the system was already known, though infrequently applied, in pre-revolutionary Russia.[11] In the years prior to the revolution, Lenin had been struck by the efficiency and scientific status of Taylorist methods.[12] He was convinced that while Taylorism was an instrument of exploitation in the capitalist mode of production, in which the bourgeoisie appropriates the surplus produced by the intensification of workers' labor, this would not be true for the socialist mode of production, which eliminated class conflict. Even if Taylor's system had "the refined cruelty of bourgeois exploitation", at the same time it encompassed "a series of very valuable scientific conquests for the analysis of mechanical motions during work, the elimination of superfluous and awkward motions, the development of more rational methods of work, and better inventory and control systems". These methods would, according to Lenin, be fundamental in the transitional phase from capitalism to socialism.[13] The increase in productivity ensured by these methods would result in a superior organization typical of socialism.

The relationship between the Taylorist system and the transitional phase that would act as a prelude to the extinction of the state, which was the ultimate goal theorized in *State and Revolution*, was repeated by Lenin in several of his works.[14] The creation of a "social regime superior to capitalism" would in fact take even longer and be more difficult than seizing power. It would require the most up-to-date technology and the material basis for Soviet industry on the one hand, and the educational and cultural development of the masses on the other.[15] Even if the lower levels of the population showed enthusiasm and a spirit of initiative, Lenin was aware that: "while we are working to increase labor productivity, at the same time we must bear in mind the particular features of the period of transition from capitalism to socialism, which require us to lay the foundations of the socialist organization of emulation, but also to use coercion to ensure the dictatorship of the proletariat will not in reality be overshadowed by the weakness of proletarian power".[16] Based partly on his reflections on Germany's war economy, Lenin strongly emphasized the need for repression and discipline.[17]

In practice, scientific management [in Russian *naučnaja organizacija truda,* NOT], quickly became an instrument in the service of War Communism in the years 1918-21 when, according to Moshe Lewin, the Soviet working class was ruined. Immediately after the revolution, the Bolsheviks had to face the problem of institutionalizing the power they had just won and guaranteeing "national survival" in the chaos of civil war. To support the war effort, they had to solve urgent production demands by reorganizing industry and agriculture, and Taylorism seemed to offer adequate solutions.[18] Along with its economic advantages, it clearly constituted a powerful tool for reestablishing from on high the discipline and control inside the factory that was needed to shore up power.

In this sense, the work of Trotsky, the People's Commissar for Military Affairs and veteran of the successful reorganization of the Red Army, was emblematic. Trotsky felt he could use his previous experience as a model to reorganize industry in post-revolutionary Russia. Thus, ignoring Lenin's clear distinction between the use of force and technology to implement organizational efficiency, he promoted and in part carried out a program for the militarization of labor that involved the imposition of military discipline in industrial relations and the total subordination of the unions to the authority of management.[19]

The Kronstad revolt of March 1921 and the New Economic Policy (NEP) did not stop the spread of Taylorism in Soviet Russia. On the contrary, by the mid-20s it had become one of the countries where scientific management was most widely discussed.[20] According to the interpretation at the time, Taylorism maintained the two-fold valence it already had in Lenin's thinking: on the one hand, it was a scientific instrument for increasing labor productivity and improving workers' standard of living; on the other, it was a means for disciplining the masses until the time came when they could discipline themselves.

NOT study centers multiplied, and with them opportunities for debate. "Bourgeois specialists", both Russian and otherwise, were called into factories to carry out rationalization, at times creating conflicts with workers. Taylorist techniques were even introduced in the planning departments, using Gantt diagrams to design the First Five-Year Plan.[21] Meanwhile, everyone continued to reflect on what form the organization of work should take in order to create a truly socialist society, one that had the workers' well-being as its top priority. Time and motion studies were carried out, aimed not so much at speeding-up movements as lessening fatigue, while the issue of improving education led to the creation of workers' schools.[22]

In the USSR, engineers were also developing a form of "humanitarian engineering" that aimed to make Taylorism superfluous by raising workers' consciousness, thereby making it possible to eliminate the original form of Taylorism envisioned for unskilled workers. On the contrary, educated workers would become the masters of their work, not the slaves, and would consequently be more dedicated.[23] These ideals were held by technicians who believed in socialist economic growth and in the prospect of socialist industrialization developed according to rational principles and long-term planning. Since this technological project was associated with Bukharin, it conflicted with the existing power.[24] The clash with Stalin took place at the same time as the launch of the First Five-Year Plan (1929-33), which inaugurated the policies of forced industrialization and of socialism in one country, as well as the series of "social wars" that shook Russia in the 1930s.[25]

The first of these wars (1928-31) was waged against the rationalization movement and the "bourgeois specialists". It led to the elimination and neutralization of most former managers and technicians who were not Communists or were at any rate skeptical about the manner of forced industrialization. They were replaced with a new technical *intelligentsia* in part made up of young super-specialists, whose education was limited to the technical aspects of engineering and the most orthodox version of Marxism-Leninism, and who posed no threat to the power of enterprise directors.[26]

The debate raised in the years of war Communism between the idea of collective management connected to workers' control and the principle that within an enterprise, power should be entrusted to a single enterprise director, *edinonachalie* [one-man command], ended in 1929.[27] The die was cast for an authoritarian model, in which the director became an autocratic boss. After 1935 collective bargaining was also abolished.[28] The Bolshevik infatuation with scientific management had been transformed into Stalinist techniques: in the USSR, Taylorism, no longer bridled by mechanisms of political control, took on terrible and grotesque forms that perverted its essential nature.[29]

The "second social war" was waged inside the factories and against the workers. Piecework, abolished by the revolution but reintroduced by Trotsky, became the most widespread method for calculating wages. In 1932 draconian measures were passed against absenteeism, the high turnover rate, and violation of internal rules, while socialist emulation was used as a forceful instrument of propaganda and consensus.[30] Until after Stalin's death in 1953, no serious large-scale discussion of enterprise management and organization took place in the Soviet Union.

Workshops and Top-Down Rationalization

The organizational work methods that had governed the phase of Soviet "primitive accumulation" in 1929-37 reached Czechoslovak automobile workshops in 1949-53. As shown in the previous chapter, manufacturing conditions at AZNP, already critical at the start of the Five-Year Plan, grew worse in the following years, especially after the intervention of Soviet advisors.

The planners' project was clearly defined: in order to be efficient, centralized planning required vertically integrated, multi-facility enterprises. Yet in order to lay the basis for mass production, it was also necessary to rationalize the organization of work. At the outset of the 1950s, despite steps to introduce Taylorist methods, pre-Taylorist work practices were still dominant in some Mladá Boleslav shops; this was demonstrated by an excessive need for skilled workers, scarce mechanization, and frequent recourse to manual operations.

In 1946-47, the years of the Two-Year Plan, the goal to increase labor productivity had been part of a complex project envisioning both organizational and technological innovations. They were meant to be carried out gradually, in accordance with the specific requirements of the sector and the Czechoslovak automobile market. Indeed, according to Taub's project, the mechanization of plants would in turn lead to job simplification and a lower percentage of manual work in certain parts of the production process, such as transportation, body, and machine work. Mechanization would also solve the problem of an over-qualified workforce. Skilled workers would be transferred to machine-tool production, while unspecialized and unskilled labor driven from the countryside as a result of the agrarian reform planned for 1947 would be used on the line.

In contrast, the scarce resources that the Five-Year Plan allocated to automobile production separated the two components of rationalization foreseen by Taub, i.e., the introduction of modern technology with single-purpose machines as well as Taylorist practices in work organization. Manufacturing equipment was not renovated, while the pressure to reduce production costs and achieve the Plan's targets intensified. In this context, Soviet-style rationalization consolidated old pre-Taylorist working practices, since it imposed quantitative objectives that could only be achieved through intensive exploitation of the workforce and machine tools as well as through substantial, informal self-management of the shops (as in the Russia of the 1930s).

Considering the existence of a serious problem of labor scarcity, the "new" methods of organization of work made intensive use of manpower an essential condition for production. The reason for the scarcity of workers, especially skilled workers, recorded in most of Škoda's technical reports from 1946 on, were partly traceable to the demographic changes in the years immediately after the war.[31] The causes of the population decline were various: the considerable military losses, the deportations carried out in the final phase of Nazi occupation, the expulsion of German and Hungarian minorities following the Beneš Decrees, and finally, the transfer of skilled workers from their original plants to others considered more strategic.[32] The scarce data available confirm a dramatic fall in the workforce employed in Mladá Boleslav alone between 1943 and 1946. Corresponding to the worst phase of Nazi deportations, the destruction of the factory, and the end of the war, the workforce had almost halved, dropping from 9,258 employees to 5,167. Afterwards, this figure at first stabilized and then grew slowly, especially after the mid-50s, due to the agrarian reform and the renewed importance of the automobile industry.

It is quite likely, however, that the labor scarcity was initially compensated by an increasing presence of women workers in factories: taking into consideration all of AZNP's employees, female labor grew from 15 percent in 1947 to 16.5 percent in 1948, and in 1950 it represented 26 percent of the personnel in Mladá Boleslav's shops.[33] What is more, the employment of women in factories was destined to grow even further, since it was encouraged by the authorities, as the 1954 issue of the magazine *Vlasta,* dedicated entirely to Škoda's female workers, demonstrates.[34] At Mladá Boleslav, women were employed for all kinds of jobs, from those usually done by auxiliary workers to more specialized work. Women worked on the machines with the assistance of "preparers" [*seřizovač*] and in manual jobs on their own. Their number was growing in the large press shops, in machine shops, and among apprentices. In contrast, the presence of women was decidedly lower among office staff, which further confirms the fact that the labor scarcity problem was chiefly in manufacturing.

The large percentage of auxiliary workers shows that recourse to rationalizing techniques was very limited in certain parts of the production process. Indeed, estimates for 1950 indicate that only 41 percent of the workforce was direct workers. Auxiliaries performed manual work that was not directly related to the manufacture of products. They were generally assigned to transport inside the factory and to inspection and maintenance jobs.[35]

Many employees were involved in transport. This is understandable, considering that the demand for electric or petrol-fueled forklift trucks needed to transport parts, which had created an authentic revolution in American workshops in the 1920s, had still not been met by the planners in 1949. The high cost of this equipment inhibited importing it, and despite the technicians' request, it was not possible to produce it either.[36] It is also true that in this period the presence of auxiliary workers was a typical feature of Soviet factories. But Soviet plants differed from Mladá Boleslav in that they could count on the availability of a vast amount of unskilled labor and on the continuous flow of peasants from the countryside.

Quality control was also carried out with rough and ready methods rather than modern automatic precision tools. The procedures used for inspection, chiefly administrative, soaked up personnel; they were also extremely arbitrary. It should also be stressed that only a small percentage of employees (one out of twenty) had technical training equivalent to two years of industrial school. The others, one out of eight employees, were experienced workers.

In the machine shop, 52 percent of inspectors were women and 5 percent invalid. At the same time, the wages of technicians in charge of inspections were 19 percent lower than those of workers in the same shop. This difference grew to 48 percent when unskilled workers acted as inspectors. The existence inside the factory of different notions of product quality and the inaccuracy of the inspectors caused another problem, which continual ministerial interventions tried in vain to solve: the high percentage of defective parts and low-quality products.[37]

Finally, there was the staff in charge of maintenance. The machines were old and broke down frequently. Repairs took place inside the shops and had to be done by auxiliary workers.[38] Machines wore out even more quickly because they were overused to meet production targets. At Mladá Boleslav there were three shifts in 1950, to take full advantage of what little modern machinery the factory had. The first shift employed 100 percent of the workforce, 16 percent on the second, and 1 percent on the third. But later data show that the third or night shift was gradually filled.[39] The shift system had been adopted in Soviet factories in the 1930s and was later abandoned. In the U.S. it was introduced only during the Second World War, while in the U.K., strong worker opposition made its introduction impossible.[40]Another instrument that proved indispensable in meeting production targets was overtime. But in 1953 *Ministerstvo všeobecného strojírenství* (MVS) [the Ministry of General Engineering] took steps to try and limit its use, since it was considered a fraudulent way to increase wages.

Along with these more traditional instruments, there were also signs of the influence of "scientific management" on Škoda's management. At Mladá Boleslav there was a "tempo and method office", though its functions were limited. In 1950 the shop foreman [*mistr*] still held an important position, even if his responsibility was beginning to be formalized within a hierarchy defined by the enterprise director. He was responsible for his squad and managed it autonomously, though within the constraints of the production plans. He supervised the use of machinery and training of apprentices, as well as assuring that production regulations and delivery dates were respected. There was one foreman for every fifty to sixty workers on the line and one for every forty to fifty workers on machine-group manufacture. The foremen did not have their own office but were assisted by a planner, a bookkeeper, and a toolmaker.[41]

Other professional profiles had also been created: norm setters (*postupář*, they established the rated capacity of each machine or the output standards for each task), piecework experts (*úkoláři,* who worked out rates for piecework), and timekeepers. The *postupář* were similar to the time-study men in milling operations. Their job was to determine technological norms but not the piece rate (*úkol,* the volume of production set for a worker in a determined time), which was the job of the *úkoláři*, and they were mainly in charge of the machines they were familiar with in the shop.[42]

Until shortly before 1948, to qualify as a norm setter it was necessary to attend a two-year apprentice school [*mistrovská škola*] and have considerable practical experience. After that date, it sufficed to have worked in the shop and have followed a fifteen-day training course on production norms. There was a *postupář* for every 160-200 workers and a *úkolář* for every 120. Charts with illustrations of production procedures were not yet used for series production, but they were being developed. Timekeepers, who were responsible for pacing the work, entered the shop in January 1947, after a fifteen-day training course. In 1950 there was one timekeeper for every 360 workers.[43]

As regards earnings, the most common system adopted was straight piecework.[44] In 1950, 60 percent of Mladá Boleslav employees were pieceworkers, including office workers. Rates varied according to the criterion on which the norm was based – that is, whether determined by chronometric calculations or the result of a time study (the estimated time for a given motion) or, as in the majority of cases, of "experience".[45] Twenty-one percent of employees were paid according to the time worked. Unskilled labor, cleaning women, and canteen workers (4.5 percent) earned an hourly wage, while office workers received a monthly salary. Both were calculated according to their category, the former with a fixed production allowance that could not exceed 15 percent, while white-collar workers received a productivity bonus that varied from 15-25 percent of the basic wage. Further bonuses were for white-collar workers with specific skills, knowledge of foreign languages, or bookkeeping.

Finally, there was performance-related bonus pay for foremen and auxiliary workers responsible for transport and controls, amounting to 19 percent of the workforce. The bonus was calculated according to categories of work and based on budget availability, savings achieved, and to what extent they adhered to the department plan. In case of a complaint about tempo, the worker had to go to the foreman, who would then inform the person in charge. Trade unions could intervene only if damage to the worker was proved. [46] For this purpose the trade union chairman appointed representatives of the various shops, who had the dual function of controlling and helping to improve organization. But even inside the factory the supreme organ was the Communist Party committee, represented by the secretary, which guaranteed that the factory followed the Party line.[47]

Just as piecework corresponded to the principle "from each according to his capacity, to each according to his work", so work on more than one machine, "an exercise of thought that coordinates machine time and manual time", was considered not only compatible with the socialist system[48] but desirable, and they sought ways to make it as productive as possible. In the manufacture of small mechanical pieces or teeth, an AZNP worker might be on as many as six machines at once, while declaring only three, and so receiving the piece rate for three machines. There were five machines for paint spray-guns, but the piece rate specified four. The type of rate did not consider the type of machine, i.e. whether it was automatic or universal, and therefore did not allow for fitting times. For two machines the piece rate was applied at 62 percent, for three machines at 48 percent, for four at 39 percent. Significantly, no minimum time per machine was set for the various operations. The preparers, specialized workers whose function was essential in a shop where semi-universal machines were prevalent, earned performance-related pay and followed the work of ten machines at most.[49]

A shop thus organized set its own pace, which depended on the skill of its workforce and on how well they worked together. Any increase in the pace of work led to a growing proportion of defective parts and waste, a fall in product quality, and an increase in the use of spare parts. Respecting unrealistic pacing norms conflicted with technological and safety limits; vice versa, if these limits were respected, it was impossible to adhere to the norms.

The "nervousness" of a worker who worked on six machines must have been even greater than in Miklos Haraszti's description *A Worker in a Workers' State*, where the author worked on two machines. The obsession, and above all the need, to maximize earnings was the only explanation for this kind of effort.[50] This becomes clear if we consider that between 1948 and 1952 the earning power of workers fell by 8 percent, and only the existence of rationing for some indispensable foods guaranteed survival.[51]

Tighter Control and Stricter Discipline

In the First Five-Year Plan, cost reduction was based on an increase in productivity. In AZNP this often translated solely into an effort to intensify labor performance. The Plan involved speeding up work tempos and taking advantage of existing resources, rather than eliminating the disorganization that reigned in the shops. The result was that existing problems became even worse, leading the authorities to adopt extreme measures. By the end of 1952 the emerging data on absenteeism, high turnover, worker-management conflicts, low product quality, and high percentages of waste and production losses were becoming alarming. The frequency of these problems was blamed on a decline in the "work morale" which posed a threat to the factory's political integrity. Consequently, morale had to be bolstered.

In the year preceding the monetary reform of June 1953, the push towards "rationalization" grew more and more intense. It took two directions – persuasion and ideological incentives on the one hand, and stricter disciplinary rules on the other. To enforce these rules, the government issued an increasing number of directives, as follows:

1. The strict definition of technical norms and standards (1951-52);
2. Administrative rationalization: introduction of a "dispatching system" and budget autonomy for each department [*chozraschot*] (1952);
3. Disciplinary measures against overtime, absenteeism, and excessive staff turnover;

4. Monetary reform (June-July 1953).

At the same time as encouraging socialist emulation, the traditional roles and behavior of skilled workers and technical managers came under attack. Along with economic reasons linked to fulfilling production plans and reducing per-unit costs, it is not hard for us to recognize how similar these measures were to those typical of the "second social war" and the struggle against functional organization [*funktsionalka*] in Soviet industry, political objectives aimed at disciplining a workforce whose support for the Communist Party was growing weaker and risked becoming problematic. In fact, these measures took on the features of a genuine "Sovietization" of the Czechoslovak industry.

The measures were in some cases issued by the Ministry of General Engineering and in others directly by the Presidency of the Central Committee of the Communist Party. They might have signified real progress towards a type of labor organization suitable for the mass production of automobiles. Yet, their coercive nature and the absence of process and product technologies equally essential for mass production brought about their discredit and transformation into an increasingly "administrative" power far removed from the interests and needs of production and incomprehensible on the shop floor.[52] This was exercising power to channel the Party's will through to the working class in "administrative practices" borrowed from scientific management, practices that were inflexible, furnished with paper evidence, and at times entrusted to the automation of machines. In this sense, AZNP at Mladá Boleslav, probably like many other production units where knowledge of original Taylorism, traces of Americanism, pride in craftsmanship and manual skills, a social-democratic tradition, and "a critical working-class consciousness" coexisted, threatened to pose a problem for Gottwald's regime, while for the Soviet Union it could be fertile terrain for insubordination and deviationism.[53]

The first blow was struck against overtime.[54] Due the scarcity of available resources, overtime had increased in order to meet the demands of the First Five-Year Plan, since without it, they could not possibly maintain scheduled output. For the Ministry, however, this was a sign of poor work organization and a means of obtaining a "fraudulent increase in wages" at the expense of the cost reductions the Plan aimed at. With the declared aim of reinstalling "economic efficiency", the regulation of January 30, 1953, guaranteed "to the organs responsible for wage policy, direct power over the management of overtime", thereby transferring this function from the office of labor inspectors and trade union headquarters to the Ministry. To this end, a series of highly detailed administrative procedures were set out in writing; they made obtaining permission for overtime extremely complicated, even when it was clearly necessary. What the majority of these measures had in common was the intention to pinpoint responsibility and put down on paper, concretely and formally, every single decision to do with even the most trivial production phase.

On April 27, 1953, the focus of the Central Committee of the Communist Party and the government shifted to defective parts and the low quality of Czechoslovak products.[55] This was another problem that had become critical during the first years of the Plan. The authorities identified in Czechoslovak factories a "lack of technological discipline" on the part of workers and technicians, who in their judgment could not be depended on to do their work correctly. Personnel intervened to modify projects and technological procedures to suit their own ideas. Consequently, scheduled production times were not respected. What is more, quality control was imprecise at all levels.

This attack was explicitly directed at the technical organs of the ministries and factories which allowed products to be made using outdated or virtually untried manufacturing methods, or technological procedures that failed to guarantee high product quality or labor productivity, and which did not contribute to setting new national standards or technical norms. The invitation to respect technical standards and norms was accompanied by a definition of clear and unequivocal responsibilities inside the factory, backed by sanctions in case of violations of "technological discipline". A high rate of defective products, hiding or disguising their presence or incomplete or sub-standard parts were equated by the court to "a crime that damages the property of the state and the nation and puts the enterprise's performance at risk".[56] A similar treatment was applied for crimes such as the delivery of low-quality or incomplete products or ones that failed to meet national standards and for damage done to products during stockpiling, warehousing, or shipping operations, which created "[...] to (our) economy the same damage as saboteurs, and for this reason factory managers, chief engineers and those responsible for technical control will be punished by the criminal court for causing damage to national property and creating difficulties for national enterprise and the fulfillment of the unified economic plan".[57]

But there was also a certain amount of confusion here. Undoubtedly, the definition of technological procedures and standardization of parts were indispensible for modernization and for mass production. Yet, the standardization had to follow Soviet norms and represented one of the most important "sovietizing" directives issued by the Bureau of the Comecon to the Czechoslovak government.[58] The Czechoslovak Association of Standardization, whose activity was limited to the mechanical engineering and steel industries, had been created in the period between the two world wars. At that time, participation in this association and respect for its standards were voluntary. Fearing competition, businesses did not join willingly and tried to meet standards only for small, simple parts. According to the authorities, this was the reason why standardization had proceeded slowly in inter-war Czechoslovakia. From this point of view, the Nazi occupation represented a step forward: the automobile industry progressed from 166 registered technical norms to 361, which followed German standards. But the foundations for a more efficient system of standardization were not laid until nationalization, since it was no longer a matter for competing private companies, but one that concerned the state.[59] Planners were also aware that especially in the absence of a market economy, it was technical norms that guaranteed product quality and were therefore the key for the planned mass production. This was true both for technical standards and for standards regulating names and definitions.

However, apart from the rhetoric, the abundant legislation in this field after February 1948 (Decree 45/51), along with the creation of a central organ for standardization, which was followed by the institution of departments for standardization in nationalized enterprises, made Czechoslovak standards fall in line with Soviet ones. Between 1952 and 1955, 301 state standards were registered for the automobile industry alone, not counting an additional set for individual enterprises; most of these measures in fact represented the introduction of equivalent Soviet standards into Czechoslovak factories.

The Mladá Boleslav plant found itself in a very difficult position indeed: forced to come to terms with continual changes in production plans, with a chronic lack of raw materials, manpower, and machinery and too few production and assembly lines, the plant found it impossible to adhere to the norms and standards set for technological procedures. Technicians and workers carried out the tasks as well as they could in the face of supply shortages of every sort, to which the administration was stubbornly indifferent. Inevitably, the empirical methods characteristic of small series manufacturing prevailed.

These measures could only be understood if seen from the rigidly ideological perspective of socialist planners, totally removed from the real conditions of production. Still, the gap between precept and reality shows how political and technological discipline joined hands in the attempt to deprive workers of an autonomy considered irreconcilable "with an iron discipline during work, with absolute submission during work to the will of a single individual, the Soviet director". The problem was that this idea of the "socialist worker" had more to do with the Russian peasant-factory worker of the 1930s than with the skilled Czechoslovak worker, and even less with technical experts unwilling to passively accept the

Soviet management system when it failed to honor their professional pride and the real conditions of Czechoslovak factories.[60]

In order to apply the ministerial measures inside the plants, the authorities increased the number of administrative tools to control production. At Mladá Boleslav a technical controls department (OTK, in Russian *Otdel techniceskovo kontrola*) was created, which answered directly to the enterprise director. This was meant to represent a step forward from the problems of quality control as described in 1950. It was an independent department whose job was to ensure that products met "national standards, technical norms, drawings and technological procedures" at every level of production, from raw materials and fuel to the finished product. Nothing could leave the factory without documentation issued by this office. Its job was to write up, mark down, and record every single procedure. The importance of putting things down in writing was meant to guarantee efficiency.

The OTK director at Mladá Boleslav was part of the production unit but had close ties to the central administration. He was appointed by the Minister on the instigation of the enterprise director, the central administration, and the director of the central product quality control office. All employees under the OTK answered exclusively to the director of this service who, by communicating to the enterprise director, had the power to interrupt the supply or shipping of products judged sub-standard, to halt production, or to demand disciplinary sanctions and bonus reductions for workers or foremen who failed to follow technical norms. In return, he was the only one responsible for quality, and as such he answered directly to the Minister.[61]

In 1952 two further Soviet-inspired organizational changes were introduced at AZNP: the dispatching system and the *chozraschot*, which corresponded to the need to create administrative instruments that would assure maximum efficiency.[62] In this connection it is interesting to note that the introduction of the dispatching system in Soviet industry in the 1930s, considered one of the chief indicators of Stalin's "totalitarian intentions", has been used by Yves Cohen to demonstrate the political valence of administrative practices "in their materiality".[63]

The *dispečerská služba* [dispatching service] was a technique derived from Taylorism for distributing and coordinating tasks. Created in the U.S. at the end of the nineteenth century, it was used in the electricity and railroad industries. The Bolsheviks were attracted by the fact that the system had been devised to coordinate the implementation of a plan. It involved analyzing the various operations and distributing them to the shops according to the availability of machines and workers. The service was made up of communications centers (dispatching services); according to Mark Beissinger: "Each dispatching service was connected by telephone to several integrated assembly lines, whose activities were coordinated by the dispatcher to prevent bottlenecks."[64] It served to keep the production flow constant, to prevent and solve possible equipment breakdowns, and, above all, to manage the consequences of irregularities in supplies.[65]

Cohen reports that in the Soviet Union, dispatching was introduced experimentally in 1932-33 in the mechanics factories in the Leningrad region, at the same time as the move to eliminate the forms of functional organization which "had represented a panacea" for every problem in the 1920s. However, it became widespread in Soviet industry only in 1937, at the height of the Stalinist terror, when it was introduced in all branches of the People's Commissariat for Heavy Industry. It is telling that in the Soviet Union, the officials in charge of dispatching were the young, hyper-specialized engineers who had replaced those eliminated in the purges against the rationalization movement. They came out of the central administration, where they were recruited, and it was their job to strengthen "opérationelle" as opposed to "fonctionelle" management.[66]

What in American industry had been a mere "technique", one of a series of arrangements, had been transformed in the USSR into a tool to eliminate "*le commandement latéral des ateliers*" and impose a "*subordination clairement verticale: les ordres tombent en cascade et la discipline est heureusement renforcée*".[67] Significantly, dispatching met resistance inside the production units and turned into a constant source of conflict between coordinators and technical personnel. Like many other measures that served to institute terror, in the end it disintegrated within the chaos it had been meant to end.[68]

In Czechoslovakia, AZNP was a pilot enterprise for dispatching. In some ways the organizational change took place unexpectedly, but at the same time very cautiously: "It met the rules of Soviet industrial administration, even if in certain details it was adapted to the structure of our enterprise structure."[69] This adaptation referred above all to the utilization of cadres from inside the enterprise as dispatchers, who not only had "energy and decision-making capacity" but also experience in the shop and practice in the section they were called on to coordinate.[70] The conflict with the technical experts and the hostility with which "conservative" cadres and foremen received the dispatch system is demonstrated by the personal intervention in October 1952 of Císař, the director of AZNP,[71] to clarify the distinction of roles and, importantly, of power between workers on the shop floor and dispatching coordinators. Though the latter were far more powerful, according to the director, their job was simply to create the conditions suitable for production, and not to interfere with it.[72] But this measure was clearly a formal distinction between operational and managing organs: the dispatchers held sole responsibility for the execution of plans, and thus could interrupt production and give orders that the shops could not question.

Resistance, above all on the part of technicians, was still recorded in June 1953. However, it seems that the system had decidedly helped AZNP to emerge from the chaos which it had slid into with the Tatra plan affair and to prepare for and start up manufacture of the new model (the Spartak-Octavia S440).[73] The system was well-suited for mass production, as comments on the first results indicate: it reduced flow time, assured a continuous flow of production, and facilitated definition of responsibilities and transparency of operations. Yet, along with their praise, the director and his close collaborators could not deny the difficulties in overcoming the suspicions, if not the downright hostility, of technicians and foremen and even of the employees chosen for the coordination system, some of whom refused to leave production to become what they felt was little more than switchboard operators. Not even using workers whom the new system had made redundant on the floor made it more acceptable and less open to accusations of excessive bureaucracy. A veritable program of re-education would have been needed, regarding political questions as well.

Indeed, doubts and impatience began to emerge because of the growing bureaucracy and proliferation of paperwork the system created. This was also due to the contemporaneous introduction of *chozraschot*, a "truly socialist method of organization of work" that consisted of calculating the profit and loss for every single labor collective. This accounting system was meant to make each worker responsible for overall production performance. In the case of Mladá Boleslav, it translated into pressure on each department to achieve greater savings in terms of costs and in severer controls to guarantee strict adherence to the output norms.[74]

Those who had doubts about the new system and felt nostalgic for the manufacturing success of the Tudor (1947) and the organization of work and production at the Škoda Shops (Škodovy Závody) were referred to an ideal of socialist organization of work that could not be realized because the "objective conditions" were lacking; in any case, someone underlined: "No organization can solve the problems by itself, without the necessary conditions and premises."[75]

Figure 4.1 – Spartak 970+955 (440) výpravna – Spartak 970, Dispatching storage (Škoda Auto Muzeum)

Figure 4.2 – Spartak 970 (440) základní lakování – Spartak 970, Painting Workshop (Škoda Auto Muzeum)

A New Approach to Socialist Labor Relations

Czech historiography defines the period 1953-57 as the first of the two crises of consensus which the Czechoslovak Communist regime had to face. According to Karel Kaplan, it ended one phase in the development of the people's democracies, the phase during which the power and economic structures that formed the social foundation of the regime were constituted, and at the same time opened a new one, which was even more firmly conducted by Moscow. Tension in the factories exploded when, on June 1, 1953, the government finally proclaimed the monetary reform it had more than once denied was in the offing. The rationing system and the presence of a double market, one under price controls and the other free, had caused strong inflationary pressures that created the need for drastic government intervention. The reform stipulated that a small part of the currency possessed by each inhabitant (300 Kč) could be exchanged at the rate of 1:5, while for the remaining currency, the rate of 1:50 would be applied.[76]

The elimination of rationing determined an overall increase in the cost of living (29 percent) out of all proportion with wage increases (4.4 percent). The final attack on what was left of the "bourgeois class" in reality hit the numerous working-class families that had until then benefited from rationing, and the measure ate up any savings they might have had.[77] Two weeks before the German uprisings there were significant spontaneous demonstrations in Czechoslovakia as well: the workers took to the streets to protest against the reform and to ask the government for guarantees that their wages would be paid.[78]

The most important demonstration took place in Plzeň, where striking workers from Škodovy Závody "defenestrated" busts of Gottwald and Stalin from the occupied city hall, calling for the return of American troops (which had liberated the city from the Nazis and fought on the side of the Czech resistance) and singing the praises of Beneš.[79] The fact that the popular militia failed to intervene to put down the demonstrations, which went on for weeks, indicates how serious a breach had grown up between the party and the sector of the working class it had been able to count on until then.[80] Busy as they were in the attempt to somehow get prices under control, the authorities reacted with extreme harshness towards the demonstrators, who were tried as traitors to the class struggle and sentenced to prison, forced relocation, and expulsion from the Party or the trade union.[81] But the sign of the Party's all-out offensive against shops that threatened to get out of control was the approval, only days after the reform, of a decree instituting penal sanctions against absentee workers and a prohibition against interrupting employment. However, this decree was very short-lived because it provoked such extreme malcontent when it was published in the Official Gazette on June 3, 1953, that only a few days later (July 7) it was withdrawn at the request of the unions, which had been encouraged by the government itself to make this request.[82]

There are no demonstrations at Mladá Boleslav on record. However, we can reconstruct the effects of monetary reform on the life of this state enterprise and observe how it related to the line taken up to then by Czechoslovak authorities in their attempts to cope with the deformations and imbalances caused by planned development in the previous five years. The wage guidelines imposed by the government and the Party's resolution on monetary reform and the elimination of rationing make it clear that though the reform was supposed to lead to a sharp cut in the price of consumer goods, in reality it turned into yet another demand to reduce running costs.[83] In order to achieve this reduction, labor productivity had to grow more than wage increases. The reduction in retail prices was supposed to restore the balance between real wages and buying power. A later ordinance affirmed that the reform had eliminated certain obstacles, laying the foundation for an increase in production and growth of labor productivity, and thereby bringing closer the time when the material and cultural levels of the population would finally improve.[84] In everyday reality, it all depended on how the state economic apparatus would "fix" the results of the monetary reform. As it turned out, the measure intensified the tendency towards even stricter discipline and tighter control in the factories.

The director of AZNP faithfully handed down to each department head the tasks assigned by the Ministry to each single part of the state apparatus. The orders were clear regarding production: do not interrupt implementation of the plan, especially where supplies for the USSR and China were in production; reduce the costs envisioned for the second semester of 1953; guarantee respect for Soviet technical norms and standards, above all for consumer goods; as regards finances: do not give budgetary credits or allow sums of money saved up to be spent; limit overtime and cut all extra spending. Significantly enough, even investments were blocked, and so was the expansion of the administrative apparatus. Enterprises also had to be prepared to meet the most important offensive: requests for pay raises that might be met by attempts to avoid or relax the rules, or to falsify statements of hours or items produced, or by applications for more production bonuses or promotions. For this reason the Ministry imposed stricter controls and forbade any increase in the wage fund at the disposal of enterprises and departments.

On the other hand, enterprises would be able to take advantage of the new manpower that would become available after the monetary reform, i.e. women and retired people. The Ministry of Labor encouraged their part-time employment, especially in light and processing industries, but they were also considered suitable for other types of manufacture. The profound cynicism underlying this truly draconian measure is evident: awareness of the impoverishment that the reform would cause to retired people and to families was associated with the realization that their need for earnings would help meet the need for industrial workers.

The part of the ordinance that dealt with socialist emulation and the effort to ensure workers' discipline "in all state and economic organs, starting from with ministries and enterprises, local municipalities and socialist farming cooperatives" was highly rhetorical.[85] Praise was showered on the enthusiasm shown by many workers in introducing the new forms of work (Soviet methods going back to the *udarnici* movement). The ordinance stressed how the monetary reform opened new prospects for the mobilization of the state economic apparatus and promoted greater "political and professional quality" of work, as well as increased individual responsibility in the struggle against bureaucratic excesses. On August 29, 1953, the director of AZNP asked his department heads to respect the following rules:

1. Describe and break down tasks for all workers, down to the least important one, and set a precise time for each task;
2. Carry out a daily systematic check of all tasks while they are being performed;
3. Assure the fulfillment of tasks while they are being performed, based on real, concrete direction of work;
4. Draw the due consequences from tasks not carried out, even to the point of substituting employees;
5. Increase your own political and professional competence and take responsibility for that of employees by setting up individual training plans;
6. Supervise the work of the personnel office and assign workers to tasks on the basis of their professional and political competence;
7. In your daily work always use the rule of criticism both for yourself and other workers.[86]

Impoverished by the monetary reform, chained to their jobs, and threatened by the penal courts, the workers had no choice but to obey and buckle down to this Soviet version of scientific management.[87]

In the interpretation of Mark Kramer, the German strikes and the Bulgarian revolts along with the Czechoslovak events of July 1953 convinced Moscow of the need not only to allow but actually incentivize reforms in east Central Europe.[88] It was Soviet intervention that mitigated the "hard line" initially followed by the new Czechoslovak president, Antonín Zápotocký, who had taken Gottwald's place after his death in 1953, and who was planning mass purges.[89]

After the events of July, Jaromír Dolanský and Viliam Široký, assisted by a group of experts, compiled the "Theses on the Present Economic and Political Situation in the Czechoslovak Republic and on the Tasks of the Party and the Government", better known as the 'August Theses'. This represented an accurate analysis of the state of the Czechoslovak economy, accompanied by a critique of the Party's political failures. Though it was decided not to discuss the Theses in public and though, acting on Moscow's advice, its harshest aspects (especially those regarding the political context) were toned down, the Theses inaugurated a season of debate on economic and organizational questions within the CCP. This debate reached its peak in Novotný's speech of June 1954 at the Tenth Party Congress, which amounted to a forthright accusation of the politics followed up to that moment.

The climate characterizing the New Czechoslovak Course also had repercussions on AZNP: in the autumn of 1953, the supply officer was requested by the Ministry of General Engineering to compile a secret document in which all the problems in the past between the production unit and the central administration would be clarified. The Minister asked for "constructive" and well-documented criticisms, as well as concrete solutions.[90]

Figure 4.3 – Octavia 993 z dia – 993 Octavia per day (Škoda Auto Muzeum)

The report was compiled in collaboration with the Factory Council, the Youth Union of Mladá Boleslav, and local representatives of the union and the Party; it was handed over directly to the Minister in November of the same year.[91] It strongly stressed the need to decentralize decision-making power and to define new relational modes between the production unit and the central administration.[92] Its judgment on the preceding period was very harsh: the authors did not miss a chance to show how a "near-sighted management" whose only real concern was following the rules had created more problems for the unit than it had solved. As for possible solutions, the main issue was how to maintain state control over the production unit without falling back into the type of overly detailed centralized planning that had shown its limitations in previous years. Finally, in a roundabout way and with extreme caution, the report dealt with the problem of the transfer of the Soviet experience to Czechoslovak industry and the need to adopt this experience with more sophistication and respect for the traditions and the existing state of production.

Resolute in tone, the report started out by expressing astonishment at the interest shown by the "supreme organ of our mechanical engineering" for the opinions of its "factories and workers". First to be targeted was the frequent reorganization that Czechoslovak industry had been subjected to over the past five years. The transition from the central administration of ČZAL to ČZAV and inclusion under the Ministry of General Engineering were considered to be disasters that had failed to create the "direct contact" between enterprise and state apparatus that was their declared purpose. In fact, the chaos generated by these measures had often obstructed the management of the enterprise. There had been countless cases of failed coordination among sectors, of frequently contradictory overlapping orders, of confusion and lack of clear priority-setting among the various branches of the Ministry.

The report went on to criticize the absence of a far-reaching vision on the part of single ministerial departments that showed up in disastrous mismanagement of subcontracting, supplies, and coordination among state enterprises and plants. Instead of resorting to efficient "coordination and internal organization", it claimed that the Ministry had merely issued orders, effectively moving more and more from a planned to a direct economy: "Operational autonomy and development of workers' initiative in the factory are more and more constrained by the recourse to orders and rules, which often interfere with even the most minute details. This management system discourages decision-making autonomy, and enterprises take advantage of this with the tactic of waiting to hear what the Ministry has to say."

Similarly, the proliferation of rules offered an excuse not to take on responsibilities or risks by making decisions autonomously. For this reason, in the rough draft the editor of the note expressed his hope that the Soviet rule of a single responsible director – one-man command – would be applied, which would mean that every worker would have only one superior whose orders he had to follow. The hierarchy would be established as follows: enterprise director, central administration director, Minister. Moreover, having access to more information about production, the intermediate director would be in a better position to respond to the factories' queries. In this way the complicated relations with the powerful, many-tentacled Ministry of General Engineering would come to an end.

A further complaint involved the proliferation of useless paperwork, which was useless not only because it was often duplicated, incomplete, and excessive, but also because no one replied to the questions posed by factories in their letters: "What does the Ministry's help consist of? They order the enterprise to get the material [...]. No advice, no help, no intervention of any sort at all, just an order."[93]

Paradoxically, to find a way out of this situation, the authors evoked the "Problems of Leninism" which had become the fundamental text of Stalinism, and a return to authentic "Soviet management".[94] What was needed was to abolish "paper management" and return to a focus on the problems of production and technology.

Another issue addressed in the report was the transfer of the Soviet experience, which had taken place too hurriedly and in the form of "orders and rules". No consideration had been given as to whether companies actually possessed what was needed for the introduction of this or that measure. The orders given from on high were never truly relevant to real conditions; they were too detailed and unrealistic, and therefore their implementation was inevitably approximate and vague. Even the rules for piecework had not been calculated properly: rates for manual workers were too low, so to earn an average wage that corresponded to an average living standard, it was continually necessary to surpass them. The gap between working time (real – needed for production) and regulation time (paid) caused a series of further inaccuracies and difficulties, making it urgent to update rates and set machine times more clearly. The consequence of the incapacity of the few was the growing disaffection of the workers for socialism. This had to be remedied as quickly as possible by bringing back the principles and values of socialist organization that had clearly been betrayed in recent years. Responsibility for this betrayal and for the difficulties encountered was firmly "returned to sender" by the production unit.

The authors of the report stressed the importance ascribed to managerial, organizational, and technical competence in "The Problems of Leninism". They observed that to assure greater efficiency, training courses were needed, including courses on economic policy for new managers and middle managers lacking on-the-job experience and management know-how, as well as training for shop foremen and workers who needed to develop their professional and political competence. Finally, what had led the Czechoslovak industry to become one of the last in the world to introduce continuous production were the many changes in production programs (four in two years), which the production unit had vehemently opposed. Mladá Boleslav had always followed the goal of mass production now proposed for the automobile industry, despite the Ministry's hostile interventions. But this goal now required new technology and a long-term, large-volume production model.

Two things are evident in this document. On the one hand, its cogency and clear understanding of the problems in the planned management of the economy allowed a debate to come out into the open that had been going on in the shops for some time, albeit in a contorted Stalinist style. While it seems to promise completely new horizons for reform, at the same time it shows evident ideological confusion as well as the impossibility of turning criticism of so-called "distortions" into a broader and more profound political analysis.

We should not overestimate the real significance of the economic changes that characterized the Czechoslovak New Course. In fact, it was a question of reorientation, and a temporary one at that, of economic policy within the system of a planned economy whose principles and institutions were never questioned in the least. Yet, the authorities' recognition of the necessity to find a new equilibrium for

Figure 4.4 – Octavia combi (Škoda Auto Muzeum)

the country's industrial development, to satisfy the needs of the population, especially for food, and to raise living standards, marked the end of intense militarization of the Czechoslovak economy and partially encouraged reconversion from heavy industry to the manufacture of consumer goods. The fall in retail prices brought about a calmer social and political climate, less hostile to the regime. For the country's most important car manufacturer, the commitment to raising workers' living standards and the new focus on consumer goods led to greater investments in the mechanization and automation of production processes.[95]

Chapter 5
The Socialist People's Car, 1955-64

Economic Reforms and Scientific-Technological Revolution

In contrast to what was happening in the same years in Poland and Hungary, the 'New Course' did not immediately represent a real challenge to the Communist Party's system of power in Czechoslovakia.[1] Antonín Novotný, Klement Gottwald's successor as Central Committee Secretary of the Czechoslovak Communist Party, kept a tight grip on the political situation, so much so that in the opinion of many scholars, the Czechoslovak regime remained the toughest, most resistant, and pervasive of the socialist systems in central Eastern Europe.[2] Still, from 1954 to 1955 the struggle against bureaucracy and recognition of the excessive centralism's limitations regarding economic management set in motion processes which would have important consequences: though they did not produce immediate proposals for political reform, these processes led to requests for greater efficiency and decentralization of the planned economy.[3] During the Second Five-Year Plan (1956-60, which followed the annual plans of 1954 and 1955) and the Third (1961-66, already abandoned by 1963), tension increased considerably between the central planning bodies trying to keep control at the top and the enterprises complaining about centralized planning.[4]

However, it was not until the serious economic crisis of 1963 that the Czechoslovak Academy of Science's Institute of Economics critically rethought the country's economic planning.[5] The need to partially reintroduce market mechanisms and end the antagonism between central planners and production units was soon associated with the demand for a different kind of socialism: one whose goal was not mere economic growth, but rather the rediscovery of an idea of socialism where growth served to "humanize" society while at the same time signifying a return to the country's own economic, industrial, and socialist traditions.[6]

The French scholar Muriel Blaive posed the question of why was Czechoslovakia so late in developing deeper political criticism of economic planning compared with Hungary and Poland?[7] In her opinion, the "resistance" and extreme "harshness" of the Czechoslovak Communist regime as well as the "Czech democratic tradition" are little more than myths of historiography. In reality, she claims, the Party enjoyed a higher level of consensus in those years than in the regimes of other central Eastern European countries. Describing the reasons for Czechoslovak "passivity" in 1956 and what she defines in her book as a "*non-événement*",[8] Blaive maintains that in Czechoslovakia, nationalism and Communist ideology had somehow converged: "rather than oppose Czechoslovak nationalism, Communist ideology in some ways managed to expropriate it, or better still embody it".[9] In her view, this "passivity" was motivated by the fact that for a long time the Czechoslovak population's standard of living was higher than that in Hungary or Poland.[10] The 1951 and 1953 protests in Prague and in the large factories were violent reactions to measures such as the monetary reform, which worsened the lifestyle of Czechoslovak workers. After these protests, the Novotný regime ably concluded what amounted to a social contract with the population, which Blaive refers to as *Stalenkovisme*: political passivity was offered in exchange for satisfying economic needs. Only the severe economic crisis of 1961-63 invalidated this tacit contract by showing that the government was no longer able to respect its terms. The crisis discredited the Communist authorities and opened up a season of economic and political demands that were to characterize the reforms of the late 1960s.

The history of the Czechoslovak automobile industry sheds a particular light on what happened after the scientific-technological revolution launched its rhetoric in 1955.[11] We must look deeper into the convergence between Czech nationalism and the socialist exaltation of production in order to understand whether the supposed "passivity" was due to the special hold a message of this sort might have had over the population, accompanied as it was by the stepped-up production of consumer goods. The investments and attempted reforms of the late 1950s seem to have oiled the gears of the system, giving the impression that it really could work, convincing specialists and workers and involving them in a renewed burst of enthusiasm for the construction of socialism, socialist motorization, and the "people's car". Yet at the same time the foundations were being laid for the final act of Czechoslovakia's disappointment with Soviet-style socialism.

The question of scientific and technological development was a key point debated by the Communist Party between the tenth Congress of 1954 and the eleventh of 1958. With the Second Five-Year Plan, and even more with the Third, the Party assured the population that technological development would lead to a higher mode of production.[12] Mechanization and automation of production processes would be the basis for this progress: along with the reduction of production costs, they would permit growth in labor productivity and reduce physical effort. Only in this way could the superior social order represented by socialism prevail over capitalism to create "a social block for peace" strong enough to successfully oppose western imperialism. A rise in the population's standard of living was inevitable, and the final result of all these efforts would be "catching up" or even overtaking capitalism in terms of technology and well-being.[13]

In these years the Soviet bloc emphasized the technological dimension of socialist modernity more and more frequently, starting a technological race with America along with subtler competition concerning consumption levels and the prosperity of the population.[14] In this context, the automobile came to represent the perfect "socialist artifact", to cite Raymond Stokes, the consumer good *par excellence*, a symbol of well-being that was also a product easily reconverted for military use.[15]

But in Czechoslovakia the automobile was more: for potential motorists, for factory workers, for a large part of the Communist bureaucracy, it was the national symbol of a historical system of production,[16] endowed with high quality and performance. In the early 1950s, sacrificing the sector might have been justified by the need to build socialism. However, now that socialism had been achieved, automobile manufacturing *had* to have as its outcome the long-desired and long-planned car that would finally lead to socialist motorization and to the slow but interrupted flow of goods on the assembly line that specialists had observed in America in 1946.

As we have seen, both on the shop floor and among those responsible for automobile production at the national level, already in 1953 (but we could easily go back as far as 1951), many people had strong reservations regarding the reality of what was happening in everyday work and the direction socialist "modernization" was taking. However, these reservations did not turn into acts of "resistance" or active protest against either the production program that led to the Tatra plan or the measures regulating the organization of labor introduced in 1951. Automobile specialists in particular continued to feel that sharing the Party's objectives was not contradicted by suggesting technical instruments of non-Soviet origin to achieve these objectives; for them, practices and principles borrowed from the capitalist industry represented rational solutions to problems of local industry conforming with socialist objectives.

In a recent study of the union movement and workers' protests in Czechoslovakia in the years 1948-68, Peter Heumos stresses that at the factory level, discontent was expressed mainly through continual production and organizational demands, which only rarely questioned the legitimacy of the Communist system. In his opinion, the lower number of strikes in the late 1950s was an indication that new, informal channels of agreement with the factory management had been opened and that the connection between workers and the political macro-structure was by then very weak.[17]

Yet, though the protests might be easy to deal with under informal factory-level agreements or attribute to deleterious local patriotism, ever more explicit signs of growing dissatisfaction with all forms of control were coming from the factories and planning centers. The recently published memoires of František Žalud and Petr Hrdlička, two engineers who played a fundamental role in planning and organizing Czechoslovak automobile production, describe the authors' daily lives in the 1950s and early 1960s as slow but continuous progress towards reaching professional satisfaction in the face of objective difficulties and bureaucratic complications. Though they were not members of the Party and did not share its language or its goals, the authors' enthusiasm for the chance to create a beautiful, modern, and competitive product that would show the world what Czechoslovak technicians could do was only partially curbed by the political context.[18]

After 1953, in terms of resources, Czechoslovak automobile manufacturing was no longer a story about the "ugly duckling" of socialist planning; on the contrary, it took on key importance for the nation's industry and economy. Compared to traditional historiographical periodization of Czechoslovak reforms, the history of the automobile industry shows a certain lack of synchronization: the events that determined its development until the 1970s, the construction of the new factory at Mladá Boleslav and the start-up of manufacturing the Škoda 1000 MB, took place between 1955 and 1965, ahead of modernization of the sector in all other Comecon countries. From this standpoint, within the limits of analyzing a single industrial sector, can the history of the Czechoslovak "people's car" help us to understand the country's delay in finding the "national way", the road it had searched for since 1946 and perhaps even since 1918?

Indeed, it was the scientific-technological revolution which, by supplying tools for plant modernization and appearing to open the way to socialist motorization, soon made it clear to experts in the sector that the modernization project launched by Klement Gottwald in 1948 would not allow them to use their know-how or to realize their western values, and least of all to remain the most advanced country of the Soviet bloc for long. Why did it take the technicians so long to realize this? Perhaps it is more useful to imagine a slow and painful change of direction rather than label it as "passivity", even more so because this label does not shed light on the great dynamism of the process that led to 1968 or on the often frustrated attempts of technicians and workers to make the socialist factory truly efficient, western, and "modern" as well as humane.[19]

New Resources and Shortages for the Automobile Industry

The Second Five-Year Plan set aside new resources to develop the light engineering industry and automobile manufacturing. In an effort to solve the problems that had come to the fore during the First Plan, the administration that coordinated its production branches underwent an overall reorganization. This had become especially necessary because of the imbalance caused by the measures introduced in the early 1950s, which had led to the exorbitant growth of heavy industry.[20] In 1955 the first step was taken to eliminate the Ministry of General Engineering (MVS), created by the Soviet advisors in 1951 and deemed chiefly responsible for the planning mistakes of 1951-53. Its range of responsibility was in fact very wide, including the management and control of all the enterprises involved in the manufacture of aircraft, weapons, motor vehicles, and agricultural machines. The result of this Ministry's activity had undoubtedly been negative: in 1953, enterprises under the MVS had not fulfilled the Plan's targets either for production or export quotas. Despite the fact that objectives were reduced on various occasions, in response to denouncements made in the three-monthly reports, in many cases the production of mechanical goods could still not keep up with growing demand. Neither the mechanization of agriculture nor the processes of rationalization and automation of production had made any progress, and the consequences for the country's food supply were disastrous. The planned objectives for military production had not been fully achieved either.

In October 1955 the *Ministerstvo automobilového průmyslu a zemědělských strojú*, MAP [Ministry of Motor Vehicles Industry and Agricultural Machinery] was split from the MVS, which had been renamed *Ministertstvo Strojírenství*, MS [Ministry of Engineering] in 1953 when it underwent a first reorganization to merge with the *Ministerstvo těžkeho strojírenství*, MSt [Ministry of Heavy Engineering Industry]. Creating a specific ministry for motor vehicles was the first and most significant sign of change in government policy regarding mobility and the manufacture of motor vehicles. The shift in the planners' attitude was confirmed in March 1956 at a conference for experts of automobile manufacturing held in Mladá Boleslav only a few months after the metalworkers' congress of September 1955, during which the Party denounced the engineering industry's serious backwardness and defined the new lines of government intervention.[21]

And so in January 1954 the Czechoslovak "people's car" began its journey, at least according to the headlines in *Svět Motorů*.[22] From that year the government started to show some degree of willingness to foster private motorization, along with a growing focus on winning the support of enterprises and production units.

Table 5.1 – Models of Automobiles (in units) Distributed by Mototechna on the Czechoslovak Domestic Market, 1968-69

Manufacturer and model	1968	%	1969	%
Škoda 1101/2	12,078	2.01	11,954	1.71
Škoda 1200/1	17,702	2.96	17,525	2.5
Škoda 1202	20,065	3.35	21,571	3.08
Škoda 440	145,478	24.3	146,428	20.9
Škoda 1000 MB	131,793	22.01	175,882	25.13
Škoda 1100 MB	1,433	0.24	12,852	1.83
Tatra 87	1,042	0.17	1,029	0.15
Tatra 600	1,561	0.26	1,543	0.22
Tatra 603	6,781	1.13	7,527	1.07
Pobeda	3,889	0.65	3,852	0.55
Volha	13,696	2.28	16,131	2.3
Moskvic 403, 407, 408	24,937	4.15	33,183	4.74
Gaz 69	2,207	0.37	2,249	0.32
Warszawa	1,470	0.24	2,670	0.38
IFA F 9	1,597	0.26	1,568	0.22
Wartburg	37,211	6.21	41,014	5.86
Trabant	22,147	3.7	29,928	4.27
Fiat 600	13,694	2.29	15,670	2.24
Fiat 850	3,523	0.59	5,546	0.79
Renault 4 CV	5,332	0.89	5,282	0.75
Renault R8	5,115	0.85	6,231	0.89
Renault R16	278	0.05	629	0.09
Simca A e E	4,976	0.83	4,928	0.71
Simca 1300, 1301	3,719	0.62	5,761	0.82
Ford Consul Cortina	5,141	0.86	5,680	0.81
Ford Anglia	1,039	0.17	1,024	0.15
Hillman Minx	2,068	0.35	2,042	0.29
Total	490,246	81.9	580,065	82.9
Other models	108,335	18.1	119,625	17.09
	598,581		699,690	

Source: Škoda-Auto Historical Archive (Aša), Mladá Boleslav, Automobilové závody, národní podnik (AZNP), b. 34, Český Státní statistický úřad, zprávy a rozbory, rok 1970, *Rozvoj automobilového průmyslu v ČSR*, tab. 33.

Table 5.2 – Motor Vehicle Imports to Czechoslovakia according to Model and Origin from 1960 to 1968 (in units)

Year	1960	1963	1964	1965	1966	1967	1968
Automobiles	13,278	12,864	16,497	15,339	20,966	22,468	35,359
GDR	4,834	3,683	4,151	4,330	10,857	9,925	13,016
USSR	4,079	5,683	5,995	3,895	3,065	5,262	957
France	1,516	927	1,216	1,632	1,831	2,725	3,517
Italy	1,529	953	1,706	1,714	2,825	2,799	3,901
Trucks	3,339	885	1,890	3,668	3,259	1,839	3,540
GDR	2,822	877	1,423	2,205	2,853	1,806	1,672
Hungary	505	1	301	620	400		

Source: Škoda-Auto Historical Archive (Aša), Mladá Boleslav, Automobilové závody, národní podnik (AZNP), b. 34, Český Státní statistický úřad, zprávy a rozbory, rok 1970, *Rozvoj automobilového průmyslu v ČSR*, tab. 17.

Table 5.3 – Number of Motor Vehicles Registered in Czechoslovakia per Region, 1968-69

Region	Automobiles 1968	Automobiles 1969	Trucks 1968	Trucks 1969	Motorcycles 1968	Motorcycles 1969
Prague	84,577	97,113	7,256	9,052	37,817	38,955
Central Bohemia	71,324	79,757	8,990	9,765	97,738	97,336
Southern Czech Lands	35,200	40,368	4,686	4,977	69,887	70,951
Western Czech Lands	45,526	52,545	7,237	7,346	66,963	66,881
Northern Czech, Lands	51,484	57,916	8,778	9,177	83,653	82,237
Eastern Czech Lands	65,450	73,898	8,583	9,020	114,479	112,455
Southern Moravia	77,802	89,754	12,000	13,018	146,240	147,572
Northern Moravia	70,789	84,620	10,984	11,489	102,949	102,917
Western Slovakia	44,715	57,179	11,630	12,326	83,578	82,847
Central Slovakia	28,340	35,221	8,891	9,374	52,924	56,575
Eastern Slovakia	23,374	31,319	8,596	9,020	41,781	41,666
Total	598,581	699,690	97,631	104,564	898,009	900,392
Governement use	42,183	48,065	97,432	102,888	52,057	49,692
Private use	556,398	651,625	199	1,676	845,952	851,096

Source: Škoda-Auto Historical Archive (Aša), Mladá Boleslav, Automobilové závody, národní podnik (AZNP), b. 34, Český Státní statistický úřad, zprávy a rozbory, rok 1970, *Rozvoj automobilového průmyslu v ČSR*, tab. 32.

Table 5.4 – Number of Inhabitants per Motor Vehicle in Selected Countries, 1966-68

Country	Automobiles and Buses			Trucks			Number of Inhabitants
	1966	1967	1968	1966	1967	1968	
USA	2.6	2.5	2.4	13	13	12	199,319
Canada	3.7	3.8	3.6	16	16	15	20,441
New Zeland	3.7	3.5	3.6	17	17	18	2,887
Australia	4.1	4	3.9	14	14	14	12,363
Sweden	4.3	4.2	4	57	59	57	7,869
France	5.5	5	4.8	26	27	25	49,890
Danmark	6.3	5.8	5.5	20	20	19	4,839
UK	5.9	5.5	5.2	33	33	33	55,068
FRG Berlin exc.	6	5.6	5.3	61	60	60	57,573
FRG Berlin inc.	6	5.6	5.3	61	61	60	59,873
Belgium and Lux.	6.7	6.1	5.7	38	37	36	9,920
Swisse	6.2	5.9	5.6	56	54	54	6,050
Norway	7.1	7.2	6.6	30	28	28	3,784
Netherlands	8.7	7.8	7	51	47	44	12,631
Austria	9	8.2	7.5	59	57	56	7,323
Italy	9.6	8.3	7.3	82	82	76	53,458
Finland	9.7	8.7	8.3	67	64	61	4,664
Japan	43	34	25	22	20	16	99,920
Spain	38	30	24	86	73	61	32,140
GDR	26	24	22	117	109	93	16,025
Portugal	37	33	30	107	103	102	9,440
Greece	79	70	64	131	130	116	8,716
Yugoslavia	98	79	55	200	192	173	19,958
USSR	229	217	199	60	63	62	233,180
Poland	135	101	94	174	161	153	31,994
Czechoslovakia	29.4	25.9	22.7	108.2	102.8	99.3	14,305
Total average	**23**	**22**	**21**	**86**	**83**	**80**	**3,356,000**

Source: Škoda-Auto Historical Archive (Aša), Mladá Boleslav, Automobilové závody, národní podnik (AZNP), b. 34, Český Státní statistický úřad, zprávy a rozbory, rok 1970, *Rozvoj automobilového průmyslu v ČSR*, tab. 43.

Table 5.5 – Total Number of Motor Vehicles Registered in Europe (including the USSR) in 1969

Country	Automobiles	Trucks	Buses	Total Motor Vehicles
FRG	11,676,000	991,500	40,700	12,708,200
West Berlin	369,700	33,500	2,000	405,200
France	11,155,000	2,015,000	50,000	13,220,000
Belgium and Lux.	1,806,500	270,000	8,900	2,085,400
Netherlands	2,100,000	300,000	9,700	2,409,700
Italy	8,178,500	760,900	37,200	8,976,600
EEC	35,285,700	4,370,900	148,500	39,805,100
UK	10,844,300	1,574,800	81,100	12,500,200
Norway	634,400	133,000	7,200	774,600
Sweden	2,003,300	137,000	12,200	2,152,500
Danmark	960,000	260,000	4,400	1,224,400
Swisse	1,260,400	120,600	4,900	1,385,900
Austria	1,053,300	107,400	6,400	1,167,100
Portugal	352,500	94,500	-	447,000
EFTA	17,108,200	2,427,300	116,200	19,651,700
GDR	827,000	182,800	13,300	1,023,100
Iceland	36,200	9,900	-	46,100
Ireland	327,400	48,000	2,600	378,000
Spain	1,599,500	553,400	33,000	2,185,900
Gibraltar	7,300	900	-	8,200
Malta	24,500	7,600	700	32,800
Yugoslavia	437,900	122,600	12,300	572,800
Greece	128,500	76,100	9,300	213,900
Turkey	93,300	106,600	25,300	225,200
Finland	580,700	97,100	7,700	685,500
USSR	1,250,000	4,000,000	75,000	5,325,000
Poland	380,000	240,000	30,000	650,000
Hungary	48,000	63,100	4,900	116,000
Rumania	22,000	65,000	2,500	89,500
Bulgaria	122,000	26,000	2,700	41,200
Czechoslovakia	699,690	173,999	18,160	891,849
Europe Total	**58,868,390**	**12,571,299**	**502,160**	**71,941,849**

Source: Škoda-Auto Historical Archive (Aša), Mladá Boleslav, Automobilové závody, národní podnik (AZNP), b. 34, Český Státní statistický úřad, zprávy a rozbory, rok 1970, *Rozvoj automobilového průmyslu v ČSR*, tab. 41.

Table 5.6 – Total Number of Vehicles Registered in Czechoslovakia from 1950 to 1969 (in units)

Year	Automobiles	Trucks	Buses	Motorcycles	Tractors	Total
1950	130,583	59,110	6,879	324,203	36,832	605,379
1951	129,733	54,426	6,250	345,967	36,608	624,585
1952	132,471	55,271	6,827	373,957	36,640	668,440
1953	136,474	55,079	6,533	397,908	37,490	701,376,
1954	135,269	59,064	7,245	446,738	39,050	760,237
1955	139,518	62,960	7,855	498,752	42,959	831,321
1956	141,916	61,872	7,915	499,577	44,716	837,130
1957	159,941	65,955	8,537	569,670	50,601	945,527
1958	188,926	69,768	9,061	644,580	58,494	1,075,813
1959	216,309	71,847	9,315	695,553	66,794	1,178,929
1960	247,092	74,571	9,860	759,494	77,395	1,301,115
1961	287,094	79,040	10,464	827,739	89,834	1,452,586
1962	318,454	81,285	11,558	865,024	101,864	1,569,772
1963	352,483	84,024	11,979	922,134	114,399	1,697,685
1964	381,564	85,920	12,402	970,921	123,921	1,808,435
1965	412,979	87,081	13,515	924,384	135,964	1,815,254
1966	455,820	90,759	14,601	920,907	145,433	1,888,955
1967	521,163	95,080	15,488	908,472	149,772	1,964,846
1968	598,581	97,631	16,575	898,009	138,010	2,031,513
1969	699,690	104,564	18,160	900,392	143,509	2,165,081

Source: Škoda Auto Historical Archive (Aša), Mladá Boleslav, Automobilové závody, národní podnik (AZNP), b. 34, Český Státní statistický úřad, zprávy a rozbory, rok 1970, *Rozvoj automobilového průmyslu v ČSR*, tab. 31.

At Mladá Boleslav the rhetoric of socialist technology seemed at least temporarily to provide some answers to the deep dissatisfaction that had been smoldering under the surface among workers and technicians during the First Five-Year Plan.[23] "Socialist technology" promised to introduce the mechanization and automation so long desired by technicians, who clearly saw them as symbols of modern automobile manufacture. At the same time, by ensuring an increase in the productivity of labor and a reduction in physical effort, the rhetoric promised a greater role for technicians and planners. Thus, a vision began to emerge of "the new social class" for whom the socialist people's car would be produced.[24]

It is therefore not surprising that the late 1950s saw a return of certain issues that had characterized the debate on motorization just after the war and in the period immediately before February 1948. People believed that mass motorization was about to become a reality in the countries of the newly constituted Warsaw Pact. The hope was that this would lead to effective collaboration and the international division of labor among the Comecon countries that had failed in the early 1950s. These countries expected that the rise in the population's standard of living would soon lead to a growth in the demand for cars, which would not suffer from the crisis of overproduction that planners already saw looming on the horizon in capitalist countries.

In the periodical literature there was a gradual increase in the number of articles dedicated initially to the Škoda 440 Spartak and then to the Škoda MB1100.[25] After 1957, however, a change becomes evident in the function that propaganda attributed to the automobile. The growing and increasingly open demand for motorization became an indicator of the high standard of living attained by workers in socialist Czechoslovakia.[26] An article significantly entitled "Motorists and Elections" which mentioned the "events in Hungary" reported these facts:

> while in 1937, 12,634 cars were manufactured, in 1956 there were 25,068 [...] but the most important thing is that whereas in bourgeois Czechoslovakia very few workers could afford to buy a car or motorcycle, now in February almost 600 million crowns have been deposited in banking institutions, the equivalent of 36,295 cars. This is an indication of the high standard of living and culture of our people.[27]

The articles on road safety that had dominated this periodical in prior years were now substituted by pieces illustrating how to order a car or justifying the taxes that still weighed on the automobile with the need to build roads, gas pumps, and other infrastructures.[28] In this optimistic perspective, the only thing that remained was to liberalize the motor vehicle market, as some were already predicting in 1956:

> Motorization penetrates our whole economic life and influences the population's standard of living and cultural possibilities [...] almost fifty thousand people who have deposited in state savings banks the twenty thousand crowns needed to buy a car, is high; but it is still limited by the fact that many of those who would be interested, have not even made the request because they are discouraged by the difficulty of using the voucher. The time has come for this number to increase along with growing production and possibly with a free market.[29]

In 1955 the planners started to focus on the Mladá Boleslav factory. They finally appeared to take into consideration some of the proposals the AZNP technicians had made almost ten years before. The majority of the presentations at the 1956 conference were in fact inspired by documents dating back to the Taub Affair and by technicians' reports of their 1947 journeys. Sometimes the authors were the same technicians who had taken part in Alexander Taub's project.

Between 1955 and 1957 plans were made to restructure the Czechoslovak automobile production. This was meant to be accomplished by 1962 with the creation of a new plant, to be followed in 1964 with the start-up of production of what was considered the first Czechoslovak people's car, the Škoda MB1000.[30]

Initially, a few limited measures aimed to rationalize and mechanize the existing plant, and some shops and lines were partially modified.[31] Introducing these measures meant a reorganization of the plant layout, which had developed in an unplanned way to meet unforeseen necessities and was often irrational in its design. When manufacture of the Š440 began in 1954, they attempted to facilitate continuous flow through the plant, despite the absence of large investments. In order to minimize the costs of moving loads within the factory, production was concentrated in a few of the main factory buildings. Warehouses were brought closer to the shops, and more than 1,500 machines were moved from one shop to another.

The "*Prague Projekta*" was finally introduced in June 1957. This was a project for a totally new plant destined to completely transform the one at Mladá Boleslav, which had grown uncontrollably since 1945 in response to contingent production needs. The goal of the *Prague Projekta* was to make it into a modern factory, rationally organized so as to reduce production costs and achieve volumes of production similar or at least comparable to European and world standards.[32]According to the planners, the new facility would mark the transition from batch or small series to mass production and guarantee significantly large production volumes. [33] And this did in fact lead to a considerable increase in production volumes and the nation's motorization; though still below the European average, Czechoslovak automobile production rose from 5,375 units in 1954 to 64,325 in 1962. Afterwards, it rose to 77,705 units in 1965 with the Škoda MB1000, while the motorization rate doubled from one car for every sixty-seven inhabitants in 1956 to one for every twenty-nine inhabitants in 1967.[34]

However, the implementation of the project – restructuring many shops and the mechanization and automation of production lines – was far from an easy task.[35] Little by little as the work continued, reports started to come in of delays, malfunctioning, and bottlenecks both in setting up the new lines and in the manufacture of the chassis. Work to achieve modernization proved problematic in the body and finishing shops. In fact, during his talk at a conference of automobile experts, Miroslav Vigner, responsible for constructing the new factory, remarked on the "conservatism and distrust of mechanization" and the tendency of the employees in the planning and investment departments to "transfer the organizational principles of batch production to larger series production". According to Vigner, this attitude led to investments being refused and thus slowed down the mechanization process. In his opinion, even norms and calculations of production

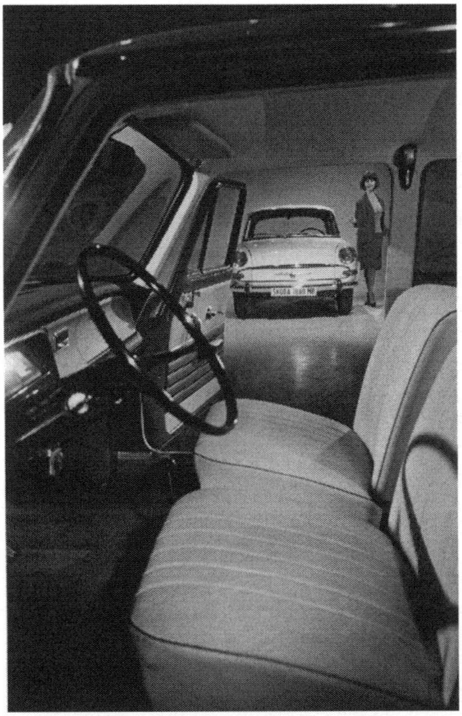

Figure 5.1 – Series Foreign Advertisement: 1000 MB interior 1000 – 1964-69 (Škoda Auto Muzeum)

Figure 5.2 -Series Foreign Advertisement: 1000 MB prospekt angl. – 1964-69 (Škoda Auto Muzeum)

Figure 5.3 – Series Foreign Advertisement: 1000 MB prospekt franc. rez autem – 1964-69 (Škoda Auto Muzeum)

capacity for plants and single machines continued to be made according to the logic of batch production.[36]

As far as the mechanization of handling material and products was concerned, the project for rationalization involved bringing in fifty-nine conveyors for a total length of about 12,500 meters. These included power and free conveyor chains, induction conveyors, belt and roller conveyors and monorails. The machines could be written-off in roughly three and a half years; bearing in mind costs, they were subdivided into three categories, according to how urgently they were required. The majority of the machines that were considered absolutely indispensable, for a total length of 5,000 meters, would be produced internally by AZNP itself. However, in Czechoslovakia serious problems hindered the development and production capacity for this kind of equipment.[37] Trouble had arisen with the hanging chain conveyors: they were produced externally, and of the nineteen conveyors required in 1955-56, only four, for a total of roughly 300 meters, were delivered to Mladá Boleslav, and very behind schedule at that.

The average AZNP machine was at least twenty years old. The special machines were still those imported from abroad in 1924-28, when "the first and last modernization of automobile production took place".[38] In this situation, there was such a great urgency for renewal that even the investments set aside to substitute machinery for the mechanization project seemed inadequate: in fact, replacement would only be completed in 1990 or 2000.[39] Still, the *Prague Projekta* envisioned significant investment in equipment, and its execution would have meant substituting general purpose machines with special ones, as well as installing comprehensive engineering units.

The number of general purpose machines in use at Mladá Boleslav was reduced from 67 percent in 1954 to 45.7 percent in 1960. In this case, too, the high productivity of the new machinery would allow it to be written-off in a very short time, which was almost trifling in the case of the automated lines, where productivity would grow by as much as 1,000 percent.[40] But in 1956-57 there were problems with the supply of transfer lines and single-purpose machines, which were considered the most efficient and useful tools for rationalizing production. The suppliers were also far from carrying out production according to existing norms and standards, and delivery could take as long as twelve months. What is more, factories refused to accept orders for single-purpose machines in excess of the production programs assigned to them by the plan. Consequently, the factory had to produce these machines itself in addition to automobile production.

The production on the automatic machining lines faced even more serious problems. A specialized factory had existed in Kuřim since 1942, *Spojených továren na obráběcí stroje Praha*, TOS [Machinery Tools Factories], but it would not be able to deliver any of the lines ordered until 1957. Thus, they considered importing some of these machines from the Federal Republic of Germany.[41]

The mechanization and rationalization project also affected other sections of the factory. Decentralization on the lines for the thermal treatment of surfaces was planned, along with the introduction of automated technical controls (equipment to measure various dimensions simultaneously, automatic equipment for measurement and selection, mechanical and electromagnetic meters). For welding operations there was to be a change-over to semi-automatic or automatic welding, with the introduction of single-purpose welders and the mechanization of fastening the pieces to be welded. Finally, they hoped to gradually introduce mechanization with assembly belts and electric or pneumatic drills. Radical measures were also envisioned for the tool-room and the body shop.

Collaboration with universities and research institutes was essential for this project. Both Bratislava's Technical University [*Slovenská technická univerzita v Bratislave,* STU] and the Czech Technical University in Prague [*České vysoké učení technické*, ČVUT] designed and created the prototypes of the machinery needed, delivering the drawings directly to the factory. Once again there was a clear need for collaboration within a coordinated framework: for example, it was stressed that the various centers that designed and developed super-finishing equipment would combine their efforts. They would work on new approaches to the treatment of surfaces, painting, and galvanization together with the Prague State Research Institute of Materials and Technology [*Výzkumný ústav materiálu a technologie v Praze*; from 1965 *Státní výzkumný ústav materiálu*, SVÚM] – at the time a center of excellence for heat treatments.[42]

But enormous difficulties arose when these operations actually got under way. They were reported in great detail and with some concern by the technicians working in the plants as well as by specialists in the relevant ministries.[43] In particular, organizing cooperation between enterprises and the various ministries responsible for automobile production proved far more complex and inefficient than expected.[44] During reconstruction work, this had created continual disputes on the quality of the goods and failure to meet delivery dates, while the efficiency of cooperation was jeopardized by long, expensive transport and high production costs.[45]

Data on the realization of the project show that progress was extremely slow and enormously difficult. This is true both for the investments planned for 1960 (they had failed to complete the plan in 1957-59) and for those following the government resolution of October 21, 1959, aimed at giving the go-ahead for construction of the new factory, which was intended to achieve a production capacity of 110,000 automobiles along with their spare parts. Despite yet another ministerial reorganization in 1962, the elimination of the Ministry of Motor Vehicles Industry and Agricultural Machinery and its merger with the Ministry of General Engineering, the problems of coordination that made it hard to find the necessary machine tools still went unsolved, and the price of these tools continued to be prohibitive. The production of the new automobile still required "a continual use of general purpose machines and equipment cheaper and more readily available than the technology of single-purpose and special machines".[46]

Numerous complaints arose from the factory about the continual increases in production planning targets while work was in progress, and the lack of negotiations to discuss these changes with the technical personnel, often considering the enterprise's tasks "unrealistic".[47]

Enterprise Reforms and Material Co-Interest of Workers

The reform was not limited to the modernization of automobile plants. The failure of the Plan and the continual problems obliged the planners to look for a new and more efficient system of enterprise management. The parliamentary debates of 1956 and the sessions of the Party's Central Committee in the following year revealed a general consensus that "management of the national economy and of individual enterprises has lagged behind the rate of development of productive forces".[48]

To solve the huge organizational problems of Soviet central planning, the government deemed it necessary to find a new system which, while leaving the state's control over enterprises wholly intact, would at the same time decentralize management and give enterprises greater autonomy and more responsibility for performance and profit. Consequently, in 1958 the planners abolished the 24 central directorates responsible to the ministries and grouped the existing enterprises into 383 multi-enterprise bodies, called *výrobní hospodářské jednotky* [production economic units or VHJ].[49]

The VHJs were of two types. Frequently, they took the form of horizontally organized "trusts", grouping similarly sized enterprises scattered over the territory and active in the same industry. In heavy engineering, the VHJs were often vertically integrated; they were given the name *concern* or *kombinat*. These units were independent from the Ministry for Legal, Budgetary and Production Matters. Their function was to mediate between enterprises and ministries, gathering and communicating production data and targets and distributing resources.

Their budgets had to be administered on the basis of cost-benefit analyses.[50] They coordinated production plans, supply of production materials, sales of finished products, and R&D for "their" enterprises.

The motor vehicles industry created twenty-five VHJs: seven trusts and eighteen branch enterprises, among them AZNP.[51] AZNP coordinated the production plans and buyer-supplier relations for its group of enterprises. Active in the automobile and component industry, some of them were quite large and included the Liberec I and Liberec II factories, the Praha Hostivař, Mariánské Lázně, Děčín, Brno, Nitra, Pieštany, Bratislava, Ružomberok, and Prešov plants, as well as the formerly independent enterprises AZ Kvasiny and AZ Vrchlabi.[52] In 1965 AZNP became the nucleus of a new trust named ČAZ [*Československé automobilové závody*]. It consequently took on full responsibility for coordinating the production, sales and servicing, and R&D for the country's entire motor vehicles sector. The trust was made up of twenty-seven national enterprises, twenty-four of which coordinated 101 factories and 80 workshops in both the Czech lands and Slovakia.[53]

The new VHJ was organized according to the Soviet principle of *chozraschot* [budgetary autonomy], and it was in charge of the industry's development, both in terms of achieving production volume targets and technical qualitative standards. Its responsibilities comprised utilization of productive capacity and plant modernization, including the construction of new buildings and production spaces. ČAZ was also in charge of subdividing and coordinating the specialization of production programs, which ensured the supply of production materials and finished and semi-finished products, while guaranteeing punctual delivery. Lastly, it also aimed to reduce costs and was responsible for training mid-level technical personnel, foremen, and workers. It seemed, therefore, that the VHJ reform at last represented the decisive stage in the process of industrial rationalization and specialization of production programs that had been pursued since the days of the First Republic.

However, the 1958 reform was in some ways ambiguous: the purpose of creating the VHJs had been to decentralize and simplify the implementation of the Plans and to eliminate the overlapping responsibilities that had paralyzed the system up to then. But it seemed that the bottlenecks and problems of the first Five-Year Plan were destined to repeat themselves, even more seriously this time because they occurred at the VHJ level rather than at the ministries. At this point, solutions to the planning uncertainties were needed at this local, less visible level, and failures would also show up locally.

The VHJ reform and its consequences have lent themselves to radically different interpretations. Alice Teichová, for example, maintains it was merely a matter

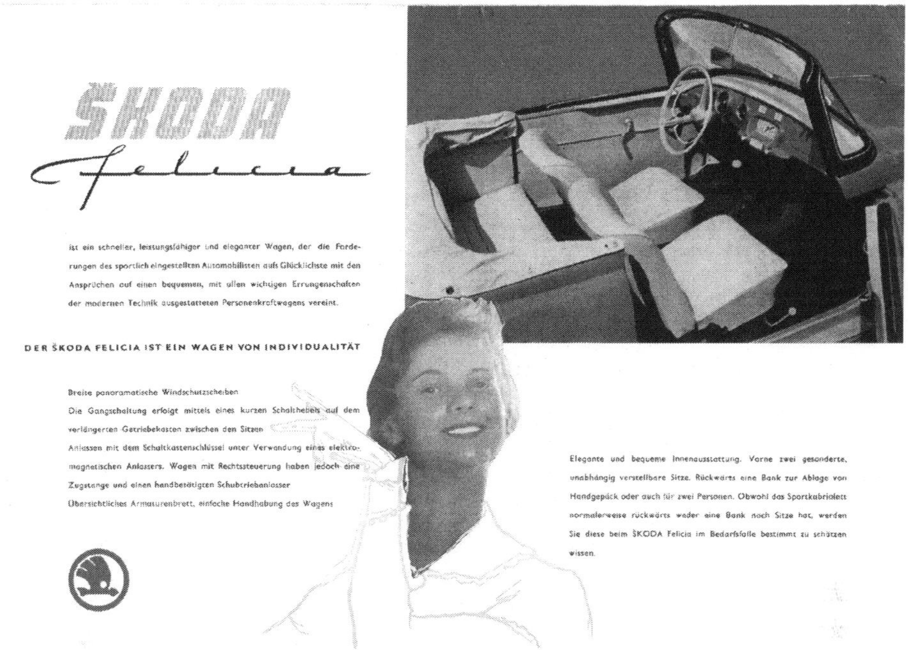

Figure 5.4 – Series Foreign Advertisement: Felicia obr. 42 – 1959-64

of trying to improve the system without really changing it by inserting a third level between ministries and enterprises: yet another – and certainly not the last – of the "twists and turns" that tried to solve the contingent problems of the planned system as they came up.[54] Others, too, comparing the 1958 reform with the later reforms of 1965-68, concur that the earlier measure served merely to aggravate the system's malfunctioning.[55]

In contrast, recent neo-institutionalist analysis has claimed that the creation of the VHJs was not a mere repetition of the periodic "twists and turns" of the Soviet management model but in fact marked the first breaking-point with that model and the beginning of the end of the political economy of Communism. The attempt to decentralize the planning system and rediscover forms of work organization more in line with Czechoslovak tradition would enable the VHJs to actuate adaptive strategies at the local level in response to the uncertainties of the "shortage economy" and the impossibility of reforming it. According to Gerald McDermott, groups of enterprises started to form local networks to access resources and protect themselves from planning malfunctions. Gradually, the VHJs and regional political institutions eroded the economic power of the center, assuming more and more responsibility for the allocation and distribution of resources. The functioning of the VHJs was entrusted to directors and representatives of the various local institutions, including Party cells, unions, and branch offices of banks, which became a sort of local control group.[56] According to this interpretation, the 1958 reform of enterprise management laid the foundations for consolidating constellations of enterprises embedded in a series of social and economic ties both vertical and horizontal at a low, informal level. These "constellations", it is claimed, were in reality far more similar to networks than the atomized and hierarchically organized enterprises envisioned by the Soviet management model.[57]

Along with the lack of adequate coordination and the difficulties in obtaining supplies and workers, low labor productivity was yet another important obstacle on the road to socialist motorization. At the same time, technical modernization provided a stimulus for adapting management techniques to the needs of the new technology and to anchor productivity norms to the truly scientific principles inherent in mass production. The First Five-Year Plan had pursued the creation of the "socialist worker" through coercion and persuasion: on the one hand, socialist emulation and Stakhanovism were encouraged; on the other, legal measures and authoritarian intervention were introduced in 1953. The results were questionable, and in 1958 the government once again decided to intervene drastically.[58] The resolution of May 30 laid the foundations for a reform of the wage system based on the "material co-interests of the workers" and entrusting its application to the various ministries.[59]

But to guarantee that the production targets set by the Communist Party's Eleventh Congress would be met – targets that as far as the light engineering industry was concerned consisted of producing durable consumer goods that would increase personal consumption – they had to find a radical solution to labor productivity. They could partly ensure this by increasing the average technical level of production and reducing the difficulty of manufacturing processes, especially for the most important products. But it was also essential to guarantee the correct use of both old and new machinery while a new product was being designed and the production cycle reorganized. As Rudolf Vykoukal said during the 1956 conference of experts held in Mladá Boleslav, the problem was to find a way to involve all the workers in the edification of socialist technology and to provide an incentive for all of them to collaborate, technicians, designers, and manual workers.[60]

At the same time it was clear that now, in contrast to what had happened with the rationalizing attempts of the First Five-Year Plan, work management systems had to pay particular attention to the special requirements of each individual enterprise: its production modes and number of lines and the optimum continuous flow of materials. So that technological innovation could produce good results, work management had to change from a model based on craftwork and batch production to one suitable for large series, continuous production. As in 1953, but using different methods, the planners were trying to solve the long-standing problem of the definition of norms, which were now "technologically correct and motivated" rather than arbitrary and empirical. For this reason, the experts felt that the question could not be handled by the wages and planning offices alone, but should involve everyone. Workers at all levels, including technicians and foremen, had to collaborate with the "*normovači*" responsible for determining production norms. But in order to meet technologically driven production norms, the technical and organizational premises on which these norms were based had to be guaranteed. The experience of the past was all too clear: if the norms did not match the technical means available and with precise technological instructions, they would not be respected, and the result would be lower profits and stagnation.[61]

The list of malfunctions compiled by the Ministry shows quite clearly that as late as 1958, the organization of work in the Czechoslovak mechanical engineering industry was still improvised and arbitrary. The frequent modifications of production programs and materials were continually changing the technological indications governing norms.[62] In many enterprises these norms were formulated without any reference whatsoever to technical instructions. This happened more

Figure 5.5 – Series Foreign Advertisement: Felicia obr. 48 – 1959-64 (Škoda Auto Muzeum)

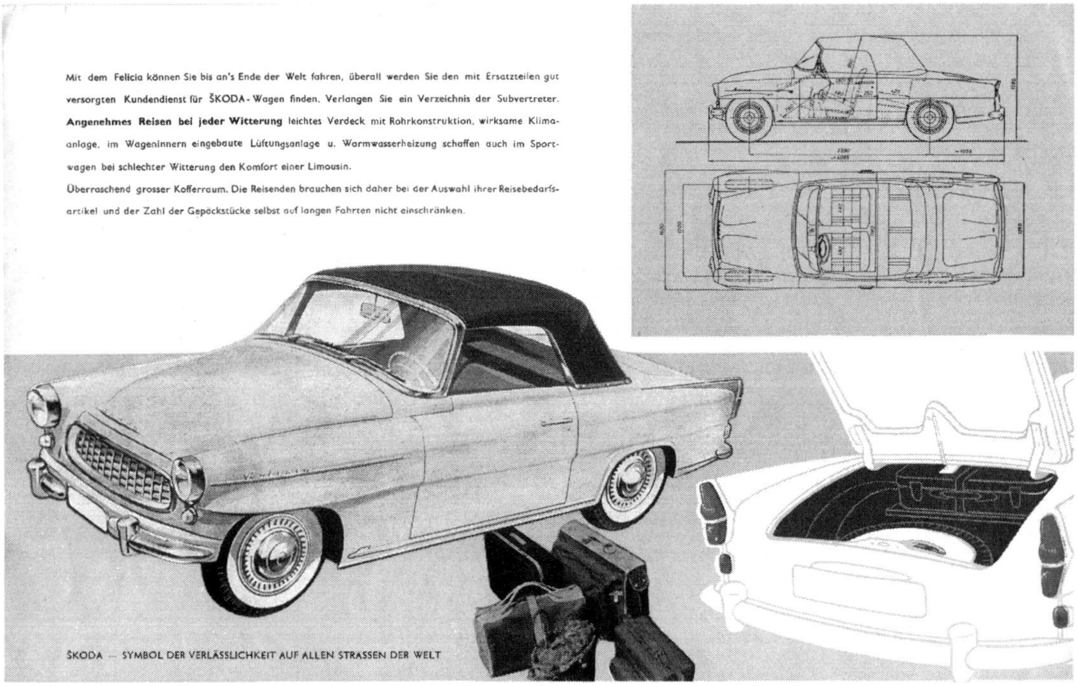

Figure 5.6 – Series Foreign Advertisement: Felicia – 1959-64 (Škoda Auto Muzeum)

frequently in metal manufacture, assembly, and generally in the phases that required greater human intervention. Norms were established according to statistical data or just experience. At times even the order of operations for a single piece of work was not defined, nor were the machines for performing each operation.

The inadequacy of norms established in this fashion had serious repercussions. On the one hand, managers considered them more or less as simple, approximate calculations of the time needed for a single operation, rather than instruments for "constant verification of the economic efficiency and performance of operations" that could improve the rational distribution of the workforce. On the other hand, the workers habitually considered any sort of definition of timing and technical norms to be a violent attack on their earning potential and not, as the authors of the document believed, a contribution to the growth of labor productivity and the "struggle against technical and organizational backwardness".[63] The authors recognized that since the workers were only paid by piecework "which barely amounts to even half of a normal income", it was natural for them to want to improve their situation by easing up the norms as much as possible: this was a way to create reserves that allowed them to earn a living wage. The technicians, on their part, were not adequately instructed about the equipment and consequently continued to give orders to work on unsuitable machines. Finally, there were the above-mentioned problems connected to planning, which ranged from delivery delays that created stoppages to the scarcity of machinery.[64]

The reform of 1958 aimed to end this chaotic situation. It had been initiated due to the need to link workers' concrete interests to promotion and an increase in work productivity. Hopefully, this would create sufficient incentive to use the new technology appropriately. There was an acknowledgement that serious mistakes had been made in the pay system, which had not motivated the workers' real interest in the quantity and quality of the goods they produced. Up to then, inaccurate linking of work to wage categories, the low quality of production norms, the failure to respect wage regulations, confusing documentation, and the inaccuracy of foremen all contributed to an imbalance in wages paid at every level – enterprises, plants, shops, and individuals. The wages of lower skilled workers ended up being equal if not higher than those of more skilled workers, and at times even than those of technicians.[65] One of the main causes of these unequal wages was the overuse of piecework compared to time-and-bonus-based wages. This had allowed some to profit unfairly and had also frequently encouraged careless work, performed without attention to product quality or safety. Among other things, this led to the overuse of tools and their consequent wear and tear. Those most responsible for this situation had been the foremen, who seemed to have misunderstood their role in the organization of work and management.[66] A comparison of various plants and sectors highlighted this lack of proportion even more.

The new wages system introduced at Mladá Boleslav on May 1, 1958, established mandatory, enterprise-wide proportions between labor productivity and average wages, introducing long-term (five-year) "norms of material co-interest". This meant that the enterprise could use its non-invested funds as an on-hand reserve to reward technical and administrative personnel, thus recreating a fair balance between their earnings and those of less qualified employees.[67]

Along with measures to define correct, hard-and-fast norms, which were to be based on technological procedures and a breakdown of work by tasks, it was considered wise to increase wage levels so that the base wage represented the major part of workers' earnings. This required resetting wage scales and ensuring a correct classification of both the tasks within wage scales and appropriate ranking. This would restore the advantages of skilled labor. The wage scales would also be in proportion to the level of social importance of the goods produced by enterprises. Piecework would not disappear, but it would be substituted where possible with time-and-bonus-based wages, either collectively or integrated with collective bonuses, to be given only if a shop or group reached specified production and cost standards. The special needs of workers involved in complex, fatiguing, or even dangerous tasks would be met in various ways, through supplements or bonuses. The workers would be motivated to conform to the new norms because there would be special pay calculated according to how quickly they complied. Finally, the new ranking and pay scales would give workers the freedom to look for jobs corresponding to their skills.

However, these were once again partial reforms. As the crisis of 1962-63 revealed, they were not able to solve the problems of production and labor productivity, which remained far lower than planners had expected until the 1970s. Yet we should not underestimate these reforms because they did indicate that the center was at least relinquishing part of its direct rigid control over work in factories: the tendency that emerged went counter to the five previous years by partially delegating management of productivity and of the workforce to the single production unit.[68]

Paradox of Exporting a Socialist People's Car

The second half of the 1950s was far from being a period of passivity. Organizational upheavals followed one upon another amid attempts to adapt various levels of production, alternating the carrot and the stick. Reconstruction of the history of Czechoslovak automobile manufacturing would not be complete without reference to another problem specialists had to face increasingly in the early 1960s, and one that might seem less important than it really was.

As stated above, in socialist Czechoslovakia, automobiles were destined mainly for export (in 1955, out of 12,430 cars manufactured, 9,441 were exported; in 1965 over 49,000 were exported out of a production of 77,700). Problems in the division of labor in Comecon and the need for hard currency had made the "traditional" markets for Škoda cars the most interesting ones for Motokov, the monopolistic enterprise in charge of automotive import-export – to Austria, Scandinavia, Belgium, and the Netherlands, all markets that worked according to the laws of capitalism.[69]

The result was a paradoxical situation: the socialist automobile was produced in a supply-centered system, viz. the "political economy of Communism", but had to be sold on capitalist markets where it was demand that counted, and at times very specific demands. Moreover, on the international market, the price had to be adapted to the product as well as to the competition. All through the 1950s

Table 5.7 – Total Export of Czechoslovak Motor Vehicles, 1948-72 (in units)

Year	Automobiles	Trucks	Buses
1948	7,074	1,435	
1949	14,611	3,338	
1950	19,380	2,921	
1951	13,504	2,480	
1952	6,762	1,587	
1953	6,167	4,554	
1954	4,776	2,692	
1955	9,441	2,527	
1956	14,718	2,486	242
1957	15,858	2,602	549
1958	25,037	4,829	593
1959	30,875	5,913	658
1960	30,556	7,362	830
1961	33,954	6,365	545
1962	34,741	6,405	616
1963	37,047	6,817	605
1964	25,419	6,047	202
1965	49,195	6,302	168
1966	51,331	6,594	75
1967	55,728	7,031	232
1968	60,527	8,179	260
1969	60,713		
1970	73,909		
1971	72,234		
1972	88,638		

Source: Škoda-Auto Historical Archive (Aša), Mladá Boleslav, Automobilové závody, národní podnik (AZNP), b. 34, Český Státní statistický úřad, zprávy a rozbory, rok 1970, *Rozvoj automobilového průmyslu v ČSR*, tab. 16, sheet 1-2.

technicians faced an indirect comparison with the capitalist industry regarding production standards and methods; but on the export markets, this comparison was direct. Starting from the late 1950s Škoda cars, both the Š440, produced since 1955, and the MB1000, the "people's car" which went into mass production in 1964, became progressively less competitive on foreign markets. By 1964 the fears expressed by technicians back in 1949-50 seemed to have come true: innumerable complaints poured into Motokov and its dealers despite the fact that Taub's factory and the Czechoslovak people's car had by then become a reality.[70] The continuity and intensity of the comparison with the products of western capitalism on the western markets shed a pitiless light on the defects of socialist technology but would have important consequences for the process of technical "learning".[71]

Between 1953 and 1967, complaints from foreign dealers, especially in Austria, Switzerland, and Scandinavia, came to the attention of the heads of Czechoslovak production and were used to various ends. The first mention of complaints from

Table 5.8 – Main Export Destinations for Czechoslovakian Motor Vehicles, 1949-55, and 1961 (AZNP – Mladá Boleslav Factory only) in units

Country	1949	1950	1951	1952	1953	1954	1955	1961
Albania		17	104	56	83		17	
Australia	462	982	890	65	72	93	148	180
Austria	296	246	263	449	551	199	610	1,163
Belgium	1,534	1,559	780	488	775	278	1,012	2,521
Brazil	490	288	460	266	1	172	222	20
Bulgaria	209	13	79	161	35	51	1	220
Canada	71	1,107	0	2	2	5	8	353
Denmark	31	82	26	344	208	185	302	1,295
GDR			3	15	25	64	106	3,963
Finland	157	308	2,205	418	1,265	1,085	1,677	4,061
France	48	77	6	6	4	1	2	291
Holland	2,200	3,142	642	217	86	121	20	1,332
Hungary	1,127	531	661	378	110	101	6	1,493
Italy	2	1	2		20			188
Norway	31			364	682	506	817	1,928
Poland	2,178	2,779	1,337	520	280	101	3	3,167
Rumania	923	718	303	213	298	1	9	80
Sweden	382	1,165	604	1,176	379	155		720
Switzerland	174	164	177	44	96	69	79	187
UK	2			6			1	1,928
USA	1						4	34
FRG	69	291	146	148	167	18	130	1,624
Yugoslavia	93						34	11
Comecon+Yugoslavia	4,530	4,058	2,487	1,343	831	318	176	8,934
Neutral Countries	1,009	1,883	3,249	2,087	2,291	1,508	2,366	6,131
Other	4,941	7,529	2,952	1,906	2,017	1,379	2,666	11,694

Source: Škoda-Auto Historical Archive (Aša), Mladá Boleslav, Automobilové závody, národní podnik (AZNP), b. 55, *Expedice vozů zahraničí od roku 1949 do r. 1955.*

foreign dealers concerning the deteriorating quality of Škoda cars was in 1953: in August and September alone, AZNP received thirty-seven letters regarding the defective quality of Škoda products, in particular the Š1200 and the T600. The complaints came mainly from Austria, Finland, Norway, and Sweden and mentioned body work, electrical equipment, various malfunctions in the gearbox, and problems with the finishing of doors and interior. Tarbruck, which distributed Škodas in Austria, had even forwarded an official complaint to the embassy in Vienna.[72] They were irrefutable complaints immediately confirmed in the strongly worded recommendations sent by Mladá Boleslav to the authorities and intended to plead the need for equipment renewal.

Table 5.9 – Main Export Destinations for Czechoslovakian Motor Vehicles, 1964-73 (AZNP – Mladá Boleslav Factory only)

Country	1964	1965	1966	1967	1968	1969	1970	1971	1972	1973
Albania			2	4			1	2		13
Australia		65	129	113	40	16	51	11	1	
Austria	305	5,503	4,077	2,632	2,551	1,769	2,787	2,329	2,137	1,537
Belgium	503	1,959	1,465	1,811	1,595	1,535	1,895	1,060	1,472	1,500
Brazil				2	1					
Bulgaria		187	1,807	1,715	1,289		200	2,105	1,237	2,006
Canada	2			2	2			40	31	1
Denmark	2	604	551	422	516	204	171	215	323	493
GDR	1,486	10,268	12,165	11,178	12,759	17,499	20,189	17,094	18,483	16,006
Finland	183	1,492	1,685	867	632	1,556	1,558	1,252	946	1,243
France	143	2,124	1,502	420	2,034	1,846	948	1,901	6,053	5,554
Holland	357	2,525	1,437	831	1,511	1,623	1,687	1,719	1,437	1,431
Hungary	2	1,243	4,271	2,350	4,056	1,955	6,309	5,566	7,357	13,265
Italy	49	334	335	406	580	616	1 175	1,736	2,177	3,001
Norway	40	1,890	948	1,191	670	1,176	633	686	481	571
Poland	1	2,495	1,597	1,426	1,804	4,764	3,027	7,048	7,105	19,584
Rumania		3	5	3,523	4,761	1,924	4,076	3,001	5,550	
Sweden	59	838	329	31	335	83	300	177	227	213
Switzerland	62	322	207	213	188	210	128	77	96	44
UK	6	1,854	1,624	1,703	362	684	1,294	2,551	8,012	11,248
USA		4			1	3		2		
FRG	207	4,994	3,319	2,136	3,362	3,266	3,422	3,398	3,302	3,504
Yugoslavia	1,052	2,521	2,081	11,632	13,836	13,117	16,137	12,550	13,458	19,963
Comecon	2,541	16,717	21,928	31,828	38,505	39,259	49,939	47,366	53,190	70,837
Neutral Countries	609	8,155	6,298	3,743	3,706	3,618	4,773	3,835	3,406	3,037
Others	1,309	16,353	11,310	9,037	10,674	10,969	11,276	13,319	23,289	27,303

Source: Škoda-Auto Historical Archive (Aša), Mladá Boleslav, Automobilové závody, národní podnik (AZNP), b. 55, *Expedice vozů zahraničí od roku 1964 do r. 1973.*

In this initial phase the central authorities held the poor work of the Mladá Boleslav workers and inefficient quality control responsible for the defects, which were consequently used to justify the introduction of socialist work discipline. But these wages and disciplinary measures did not solve the problem of defective items and rejects. On the contrary, the complaints increased even after the launch of the model Š440 in 1955, which was meant to represent a step forward.[73]

In 1958 there were 4,333 complaints in Sweden alone, concerning almost every aspect of the Š440. And as the distributors stressed, the problems could not be attributed to the particular climatic conditions in Scandinavian countries.[74]

And even when the Š440 went out of production in 1965 and the Škoda MB1000 finally came in, the truly heartfelt words spoken by the founder of the Austrian Tarbruck to dealers made it clear that not much had changed since the first official complaint dating back to 1953, and that Škoda still had a long way to go if it wanted to launch a product that would be competitive on western markets.[75]

Nevertheless, in reality something had changed: while most of the earlier complaints had been about the technical quality of Škoda cars, by the early 1960s what was contested by dealers and discussed with Czechoslovak authorities was the lack of competitiveness of Škoda products as regards esthetic appeal, marketing, service, and technical assistance abroad.[76] They took this issue very seriously in Prague, as shown by the intense exchange of letters between the Communist Party Central Committee, the production ministries, local Party cells, Motokov, and AZNP. The importance they attached to the letters of complaint and the measures taken afterwards was justified not only by the concern for commercial repercussions, but also by the fear that car defects could be blamed on acts of sabotage or political disobedience on the part of the workforce, and thus exploited for anti-Soviet propaganda.[77]

The complaints shed light on design defects in Škoda models Š1200, T600, and Š440; later on they were also useful to identify certain problems with the Škoda MB1000. Thus, these letters probably served as catalysts for the decision to invest in the modernization of automobile production in the late 1950s, especially as many of them came from friendly nations that had not cut off trade relations with the Czechoslovak manufacturer even in the years of greatest political tension. They also seem to have served to orient the measures taken to modernize the Mladá Boleslav plant, since they converged with the needs expressed by technical personnel.[78]

Motokov's growing difficulty in selling Škoda products abroad led it to contest the organization of exports as developed in the early 1950s. Gradually, they chose to give more scope to dealers in formulating techniques to improve the quality of Škoda products, along with more direct involvement of the manufacturing enterprises. To this end, Motokov drew up a series of questionnaires on the quality of Škoda products and the organization of exports, which it sent out to dealers and to Mladá Boleslav. Later on, it arranged periodic meetings at which its delegates met up with AZNP technicians as well as foreign representatives.

As for AZNP, in the late 1950s the letters of complaint came with demands from the manufacturing enterprises for renewed decentralization of the economic decision-making process, and especially of relations between exporters and manufacturers. In 1955 Mladá Boleslav's answers to a questionnaire from the ministry underscored the need to go back to the way things had been done before 1950. This meant reestablishing contact between manufacturing enterprises and foreign distributors, once again starting up the visits and the direct, almost personal relations which the experts felt were the only way "to allow us to keep in step with foreign competition, technical progress, the exchange of information and patents" and which had "positive effects from the psychological standpoint on foreign delegates and customers". Contact with dealers could not continue to be filtered through Motokov and pass only through the complaints department, nor could Mladá Boleslav continue to be the scapegoat for every failure of efficiency.

The questionnaire revealed that the lack of direct communication between production and distribution was causing problems in providing technical assistance for Škoda cars sold abroad.[79] Škoda only had service points in Belgium, Austria, and Switzerland. Mladá Boleslav stressed the importance of sending technical personnel to retailers to set up real assistance stations equipped with expert, technically competent staff from the Škoda production plant. This would be even more important if the proposed idea of organizing automobile assembly abroad, in Austria and Belgium, was accepted. Along with the obvious need to set up assembly plants, this proposal would involve creating a genuine network of foreign agents.

Still, in 1954 it was Motokov itself that sent fifteen foreign enterprises – among them: Oslo, Lucerne, Pretoria, Continental Motors Colombo, Casablanca, Teheran, Australia, Brussels, Turkey – the first questionnaire to ask for suggestions on how to improve the distribution of Škoda products and their main defects. Later on, Motokov organized meetings between its foreign delegates and AZNP.[80] The slogan of the December 1958 meeting was "the experience of foreign workers as input for the elimination of production and technical-commercial defects in the export of Czechoslovak automobiles". Its clear goal was the improvement of collaboration between Motokov, foreign dealers, and the manufacturing plants.[81]

On this occasion the Swedish delegate, referring to current market conditions in Scandinavia, reminded AZNP and Motokov that by then it was normal to change cars fairly frequently. He regretted the Czechoslovak partners' lack of attention to certain aspects, obviously pertaining to marketing. He criticized the lack of advertising campaigns and of small changes to the finish of the exterior, pointing out that the range of colors did not meet market demands and that in general the Czechoslovak enterprise hardly considered the purchasing habits of Swedes. He made it perfectly clear that the decline in Škoda sales in Scandinavia was not the fault of politics, but rather that the reputation of Škoda cars had been falling since 1953, and that this was due to their poor quality and lack of service.

We do not know how the AZNP representatives and Motokov reacted to this criticism, which was repeated in 1965 when marketing and advertising once again became the focus of discussion. But when in those years a video advertisement aimed at domestic consumers showed a suitcase-stuffed MB1000 driving along the snowy streets of the new socialist city with an attractive young lady at the wheel and a *Beatles* soundtrack, spring was already on its way to Mlada Boleslav.

Epilogue: Prague Spring 1968

In the course of the 1950s, there was a deepening gap between the technicians' knowledge acquired in the years 1946-47, which continued to be their technical terms of reference, and the actual operations in the plants, which still used old machines and methods with limited production volumes.

Though there was an attempt to impose the Soviet management system on recalcitrant factories, the technicians' reports still referred to a mixed model for overall productive efficiency, incorporating some elements of the American model, especially single-use machines and high-quality steels. This went hand in hand with admiration for the supposed achievements of Soviet industry (often not based on first-hand knowledge) and with frequent references to the Czechoslovak industrial tradition, meaning the organization and know-how gained throughout the sector's progress before World War II. Considering all these aspects, the technicians stressed the dangers of marginalizing automobile manufacture, which was not only a traditional Czechoslovak industrial product but was also a key industry for the country's balance of trade.

During the Second Five-Year Plan (1956-60) and the first phase of the Third, the Mladá Boleslav plant was transformed by massive investments aimed at matching the process technology to the requirements of mass production. This would eliminate the imbalance between technology, organizational structure, and management systems that had widened from year to year, characterizing post-war growth. It seemed that by eliminating the most glaring contradictions of the First Five-Year Plan, a new plant would be able to create the conditions needed to meet the productivity promises of the Soviet management model and of mass production.

But it was precisely when Škoda was about to become a truly large-scale enterprise specializing in automobile manufacture on a national scale, with a partially automated plant that could produce a people's car in large volumes, that the most serious problems emerged: the specialists' search for efficiency and quality came up against the difficulties of a planned economy in coordinating production as complex as manufacturing automobiles. In these years the gap between the technicians' expectations and official debates about the automobile and the "socialist driver" on the one hand, and the reality of Czechoslovak factories on the other, not only became clear but grew wider. The image of an efficient national industry was undermined, thereby sowing the seeds for criticism of a technical-productive nature underlying the movement for political reform in the 1960s.

In the end, the investments in productivity and the modernization of the Mladà Boleslav plant actually brought to light the shortages and problems facing automobile manufacture, based as it was on a multiplicity of pieces, components, and materials in a centrally planned economy.[1] The technicians who worked in the automobile sector could not fail to see that the gap now separating them from their capitalist rivals was indeed huge. Between 1955 and 1964, there were many attempts to solve the problems concerning production processes as well as the quality of the automobiles. Yet, no matter how much attention planners paid to reports from the plants and above all from Motokov, ultimately the problems posed by the requirements of efficient automobile production would challenge the very foundations of economic planning and touch on politically sensitive issues like liberalization of the domestic market, the introduction of market pricing mechanisms, and the use of marketing strategies and advertising for foreign distribution.[2]

The crisis of 1963 marked the start of real economic reform in Czechoslovakia.[3] At Škoda this meant the gradual radicalization of proposals for innovation and the reorganization of relations with the ministry, and even more so with distributors. They were concrete proposals, born of a comparison with the capitalist industry and without any particularly revolutionary features from a political or ideological standpoint. They were spurred by the need to solve existing problems and to restore efficiency and prestige to the Czechoslovak automobile. The technicians' reports clearly show that they finally felt free to call things by their real name. In solving the problem of bottlenecks that impeded a continuous flow of production, they were able to shake off the embarrassment and shame about a product that took great effort to manufacture but was hard to sell and did not adequately represent Czech know-how.

The instruments of capitalism – the market mechanism, the function of prices, the marketing strategies, the internalization of distribution and of research and development – were not an end in themselves but rather concrete solutions to managing enterprises in order to ensure productivity and efficiency. As in 1946, restoring efficiency to industry was how technicians and workers contributed to paving the Czechoslovak way to socialism. And once again, they did so by turning their attention to practices and technology used beyond the iron curtain.[4]

But in their search for modernity and efficiency, the technicians were introducing ideas that risked overstepping the confines of socialist ideology. In June 1967, a conference on marketing and its application for AZNP was held at Mladá Boleslav.[5] Its objectives were to clarify the definitions of "marketing" and "public relations", to demonstrate forms of market research and the role of advertising, to explain how to use marketing in the automobile sector, and to "build a positive image of producer and product and influence customers before they decide to buy". Considerable scope was also given to the question of pricing: "how to set suitable prices [...] how to evaluate when it is possible to set a price higher than the competition."[6]

Looking closer at the conference reports shows us that they were studying all these things: simplifying distribution "following the model of other industrially developed countries", improving the possibility to maintain Škoda cars (service and access to spare parts and repairs), and the complex mechanisms of price formation. They were seeking solutions to problems in export in order to adapt to trends on the international automobile market rather than to the pressing demand of the domestic market. Besides the need to improve the bodywork and possibly the options on offer, the most radical proposal involved Motokov's role – the need to make it an internal division of AZNP that would be in charge of all aspects of trade and distribution, not only sales but also service, advertising campaigns, and so on. At the same time, they proposed decentralizing distribution and service at the regional level for the domestic market, so that "the enterprise would transfer only a part of distribution, without having a monopoly on sales; on the contrary, it would be a competitor of Mototechna, Obchodní dům, and so on".[7]

However, this study of market strategies and innovative distribution systems was a sign that things were going too far: rejecting Sloanism and mass marketing had been an absolutely fixed concept in distinguishing between capitalist and socialist automobile industries. If as far back as the 1950s the Sloanist principle of "a car for every purse and purpose" had made sure but slow progress in capitalist Europe, behind the iron curtain cars were and had to remain "utility goods" – useful, unsophisticated, meant to teach technology and discipline, and produced according to the cannons of Soviet-style "exaggerated Fordism".[8] With some updating, the reference was still the Ford Model T: motorization would happen with one sole model, and the market would be created thanks to low selling prices, low production costs, and maximum standardization, as in 1920s America. Besides complicating the manufacturing process, the proliferation of options and models implied recognition of the consumer's individuality and the need to respond to individual preferences.[9] As the growing literature on socialist consumption shows, this sort of recognition risked taking on ideologically subversive connotations and escaping the planners' control. What is more, in Czechoslovakia an increase in the range of models "to satisfy even the most exacting motorists, those who haven't liked our cars up to now" recalled the large variety of models that had characterized the deleterious, status-symbol motorization of the 1930s. The sophistication and flair, once a trademark of

Czechoslovak cars, had been the precise object of ideological barbs throughout the 1950s.

But in 1967, potential Czechoslovak customers whose voices began to be heard and whose wishes began to count resembled even more the needs and expectations of the production world – in a long televised interview in 1967, this is how the director of AZNP answered questions concerning the future of Czechoslovak motorization and sales with installments:

> demand is higher than supply and is growing very quickly; this has translated into a growing number of orders that we haven't been able to meet, a growing number of people who want a car but haven't received one. We cannot know yet how supply will evolve; what is certain is that demand will also depend on the price of cars. Retail prices and their variations depend on the economic situation in our country; they are undoubtedly also influenced by production costs, and this is obviously the business of the producer and of relations between producer and supplier. We will try to produce in the best and most economical way, but the change in demand also depends on automobile tax, the cost of gas, the quantity, quality and price of spare parts, the state and development of the road network. Another no less important factor will be the rise in living standards, and with it the growth and differentiation of wages.[10]

A lot had changed since 1949, if not in the factories then in the language and approach to automobile production and consumption. At the same time, at Fiat's headquarters in Turin, a big marketing campaign for the Czechoslovak Republic was already being prepared: someone already imagined a giant Fiat advertisement dominating Venceslav Square. Elsewhere in Italy, Giuseppe Luraghi, Alfa Romeo's president, had already begun negotiations to produce Alfas in Czechoslovakia with a counter compensation agreement to receive the new automatic machines produced by TOS in exchange.

However, the response to the contradictory plans for developing the Czechoslovak automobile industry and to the uncertainties and hopes aroused by the Czechoslovak path to socialism arrived very soon in the shape of Soviet tanks. Mass motorization followed. After August 1968, the planners briskly pursued motorization for individual use: despite continual problems in production, for the time being Škoda found less demanding consumers in the Comecon markets. Only a few hundred kilometers away, the Volga plant, inaugurated in 1970, loomed up to symbolize the only possible way for socialist automobility: to import the most advanced technologies of western automation without yielding to the perversions of capitalist consumption in order to produce an automobile which, to quote Ford, the consumer could paint "any color that he wants so long as it is black". As in the best Fordist tradition, this would distract workers and, in the case of Czechoslovakia, citizens from commitment and participation in politics or unions.

Bibliography

Archival Sources

Archiv Akademie věd České republiky (AV ČR)
- Masaryková akademie práce (MAP)

Archiv společnosti Škoda auto (Aša)
- Akciová společnost pro automobilový průmysl (ASAP)
- Akciová společnost pro automobilový průmysl účtárna (ASAP/U)
- automobilové závody, národní podnik 1945/1948 (AZNP/P)
- automobilové závody, národní podnik (AZNP)

Státní ústřední archiv (SÚA)
- Československé závody kovodělné a strojírenské (ČZKS)
- Československé závody automobilové a letecké (ČZAL)
- Ministerstvo všeobecného strojírenství (MVS)
- Ministerstvo automobilového průmyslu a zemědělských strojů (MAP)
- Hospodářská rada, ústřední výbor KSČ (HR ÚV KSČ)
- Úřad předsednictva vlády – tajná spisovna (ÚPV-T)

Newspaper and magazine articles

Czechoslovak Economic Papers
Rosický, Miroslav. "Organization and Management of Industry and Construction in the Czechoslovak Socialist Republic." no. 2 (1962): 60–61.

Podniková Organisace
Vigner, Miroslav. "Vliv účelného uspořádání výrobního pochodu na hospodárnost." no. 7 (1955): 264-267.
Pivonka, J. "Mistr – organisátor výroby, socialistickího soutežení, rádce a vychovatel." no. 9 (1955): 350–352.

Encyklopedie Výkonnosti
Verunáč, Václav. "Člověk a hospodářský život." (1934): 604-617.

Nová práce
Hudec, Karel. "Fordův automobilismus." no. 15 (1922): 203-207.
Karel Hudec, "Henry Ford." no. 11 and 12 (1929): 149-162.

Mužík, Václav. "Z prakse ve Fordových závodech v Detroitu." no. 3 (1925): 44-45.
Suchánek, Jaroslav. "Našemu autoprůmyslu." no. 9 (1927): 74-75.
Suchánek, Jaroslav. "Fordův dělník." no.8 (1926): 168-170.
"Škodovy závody jako podnik technického a sociálního pokroku." no.7 (1925): 366.
"Ing. Francesco Mauro první president mezinárodní delegace pro vědeckou organisaci pracovní." no. 8 (1926): 4.
"Charakteristické typy Fordových Závodů." no. 9 (1927): 169-175, 185-186, 204-206.
"Použití Fordových metod u nás." no. 10 (1928): 164–165.
Mimra, Bohumir. "Je možný u nás levný vůz domací výroby?" no. 10 (1928): 162-163.

Zprávy Československého národního komitétu pro vědeckou organisaci
"Tomáš Baťa." no. 4 (1929): 67-68.
"Dr. Techn. H.c. Ing. Josef Havránek." no. 9 (1929): 183.
"Walther Rathenau." no. 6 (1930): 133.
"Výroba a provoz." no. 3 (1930): 57
"Václav Klement." no. 1 (1931): 9.
"Mario Fossati." no. 4 (1931): 109-110.
"Výroba osobních a dopravních automobilů Škoda v Mladé Boleslavi." no. 2 (1931): 17-18.

Právo lidu
"Pryč s racionalisací, prováděnou na účet dělnictva!" June 18, 1930.

Technický ozbor
Fleischner, Jindřich. "Technická kultura." no. 24 (1911).

Cement, železo a beton
Špaček, Stanislav. "O pracovní a lidské ekonomii." no. 8 (1914).

Vzdělání lidu
Špaček, Stanislav. "O lidské ekonomii." No. 9 (1914).

Obzor průmyslový
Špaček, Stanislav. "Taylorismus a jeho význam pro práci v době mírové." no. 1 (1915-16).
Špaček, Stanislav. "Taylorismus ze stanoviska pracovní a lidské ekonomie." no. 1 (1915-16).

Annals of the American Academy of Political and Social Science
Wallace, L. W. "Industrial Management and the American Engineering Council." Science in Modern Industry, no. 119 (1925): 127-133.

Svět motorů
"Svět hledá malý automobil." no. 33 (1948): 228
"Debata o lidovém voze." no. 29 (1948): 97-98
"Reorganisace Autoklubu RČS a jeho příští úkoly." no. 74 (1950): 99.
"Vývoj vědeckého výzkumu automobilů v SSSR." No. 77-78-79 (1950): 205
"Nemístní luxus." no. (1950): 164.
Tůma, Adolf. "Rozvoj motorové dopravy." no. 83 (1950): 387.
Tůma, Adolf. "Motorisace Sovětského svazu." no. 91 (1950): 643.
"Motorismus v Sovětském svazu." no. 95 (1951): 786.

"Sovětský automobilní průmysl." no. 98 (1951): 870.
"Zvýšíme brannost našeho národa." no. 116 (1951): 1443.
"Poslání a úkoly Československych motoristů." no. 6 (1952).
"Reorganisace Mototechny." no. 122 (1952): 97.
"Malý automobil - velký problém." no. 161/162 (1953).
"Naše cesta k lidovému vozu." no. 164 (1954): 4-5, continuing in no. 165 (1954): 38-39.
"Spartak-mezityp lidového automobilu." no. 189 (1954): 756-757.
"Ještě něco o Spartaku." no. 190 (1954): 53.
"Automobil Spartak v měřítku světové konkurence." no. 191(1954): 206.
"Deset let vývoje a výroby automobilů." no. 198 (1954): 298-299.
"V Itálii a u nás, na pomoc předvolebním besedám." No. 195 (1956): 742.
"Lépe využívat zahraničních technických poznatků a zintensivnit propagaci nové techniky." no. 209 (1955): 1.
"Prodej osobních automobilů v roce 1957." no. 240 (1956): 823.
"Prodej osobních automobilů obyvatelstvu v roce 1956." no. 218 (1956): 122.
"Kdy a jak si budeme moci koupit automobil." no. 244 (1957): 107–108.
"Motoristé před volbami." No. 250 (1957): 1.
Šimon, Karel. "Strana a vláda pečují o rozvoj motorových." no. 261 (1957): 1.
Jirout, Václav. "Více, lépe a levněji." no. 262 (1957).
"K některým otázkám daně z motorových vozidel." no. 266 (1957): 1.
Šimon, Karel. "Strana a vláda pečují o rozvoj motorismu." no. 261 (1957): 1.
Jirout, Václav. "Více, lépe a levněji." no. 262 (1957).

Sborník MAP
ŽMAVC, Ivan. "Jak MAP by mohla býti nápomocna k řešení úkolu národní a světové krise?" no. 1 (1932): 1-3.

Books and Periodicals

Abelshauser, Werner. "Two Kinds of Fordism: On the Differing Roles of the Auto-mobile Industry in the Development of the Two German States." In *Fordism Transformed, the Development of Production Methods in the Automobile Industry*, edited by Haruhito Shiomi and Kazuo Wada, 269-296. Oxford: Oxford University Press, 1995.
Abrams, Bradley. *The Struggle for the Soul of the Nation: Czech Culture and the Rise of Communism*. Oxford: Rowman and Littlefield Publishers, 2004.
Adler-Karlsson, Gunnar. *Western Economic Warfare 1947-1967: A Case Study in Foreign Economic Policy*. Stockholm: Almqvist & Wiksell, 1968.
Aldcroft, Derek Howard, and Steven Morewood. *Economic Change in Eastern Europe since 1918*. Aldershot: Edward Elgar, 1995.
Amatori, Franco. "Cicli produttivi, tecnologie, organizzazione del lavoro. La siderurgia a ciclo integrale dal piano autarchico alla fondazione della Italsider 1937-1961." *Ricerche Storiche* 3 (1980): 557-611.
Amatori, Franco. *Impresa e Mercato. Lancia, 1906-1969*. Bologna: Il Mulino, 1996.
Anfossi, Anna. *Prospettive Sociologiche sull'organizzazione aziendale.* Milano: Franco Angeli, 1971.
Anselmo, Marcello. "Der werktatige Verbraucher: Defining the Socialist Consumer; Market Research in GDR 1960s/ '70s." In *The Sovietization of Eastern Europe: New Perspectives on the Postwar Period*, edited by Balázs Apor, Péter Apor, and Arfon E. Rees, 77-91. Washington: New Academia Publishing, 2008.

Apor, Balázs, Péter Apor, and Arfon E. Rees, eds. *The Sovietization of Eastern Europe: New Perspectives on the Postwar Period*. Washington: New Academia Publishing, 2008.
Bailes, Kendall E. *Technology and Society under Lenin and Stalin: Origins of the Soviet Technical Intelligentsia 1917-1941*. Princeton: Princeton University Press, 1978.
Bardou, Jean Pierre, Jean-Jacques Chanaron, Patrick Fridenson, and James M. Laux. *The Automobile Revolution*. Chapel Hill: The University of North Carolina Press, 1982.
Bartošek, Karel. "Stávkujicí dělník v práci, roku 1953." *Listy* 17, no. 2 (1987)
Bašus, Albín. *Moderní organisace v strojírnách: studie technicko-hospodářská*. Prague: Nákladem vlastním, 1911.
Bazala, Alberto. *Il pensiero di Masaryk*. Praga: Biblioteca Internazionale di Filosofia, 1935.
Beaud, Charles. "Investments and Profits of the Multinational Schneider Group, 1894-1943." In *Multinational Enterprise in Historical Perspective*, edited by Alice Teichová, Maurice Lévy-Leboyer, and Helga Nussbaum, 87-102. Cambridge: Cambridge University Press, 1986.
Beissinger, Marc. *Scientific Management, Social Discipline, and Soviet Power*. London: I.B. Tauris & Co., 1988.
Benvenuti, Francesco. *Fuoco sui sabotatori. Stachanovismo e organizzazione industriale in URSS, 1934-38*. Rome: Valerio Levi, 1988.
Berliner. Joseph. *Factory and Manager in the USSR*. Cambridge: Harvard University Press, 1957.
Bettanin, Fabio. *Stalin e L'Europa. La formazione dell'impero esterno sovietico (1941-1953)*. Roma: Carocci, 2006.
Bibò, Istvan. *La miseria dei piccoli Stati dell'Europa Orientale*. Bologna: Il Mulino, 1994.
Bigazzi, Duccio. "L'ora dei tecnici": aspirazioni e progetti tra guerra e ricostruzione." In *Pensare l'Italia nuova: la cultura economica milanese tra corporativismo e ricostruzione*, edited by Giuseppe De Luca, 379-433. Milano: Franco Angeli, 1997.
Bigazzi, Duccio. "Modelli e pratiche organizzative nell'industrializzazione italiana." In *Storia d'Italia, Annali Vol. 15, L'Industria*, edited by Franco Amatori, Duccio Bigazzi, Renato Giannetti, and Luciano Segreto, 897--962. Torino: Einaudi, 1999.
Bigazzi, Duccio. "Organizzazione del lavoro e razionalizzazione nella crisi del fascismo 1942-1943." *Studi Storici* 2 (1978): 367-396.
Bigazzi, Duccio. *La Grande Fabbrica. Organizzazione industriale e modello americano alla Fiat dal Lingotto a Mirafiori*. Milano: Feltrinelli, 2000.
Blaive, Muriel. *Une déstalinisation manqué. Tchècoslovaquie 1956*, edited by Krzysztof Pomian. Paris: Editions Complexe, 2005.
Bloomfield, Jan. *Passive Revolution: Politics and the Czechoslovak Working Class 1945-1948*. London: Allison and Busby, 1979.
Boffito, Carlo, and Lisa Foa. *La crisi del modello sovietico in Cecoslovacchia*. Torino: Einaudi, 1970.
Bottoni, Stefano. "Politiche nazionali e conflitto etnico. Le minoranze ungheresi in Europa orientale (1944-1955)." *Contemporanea* 1 (2002): 85-116.
Brom, Bohumir and Eduard Kubů. "The Role of the Czechoslovak Trade with Neutral Countries in the Period of Escalating Cold War: The Example of Switzerland and Sweden 1949-1953." In *Gaps in the Iron Curtain: Economic Relations between Neutral and Socialist Countries in Cold War Europe*, eds. Alice Teichova, Gertrude Enderle-Burcel, and Piotr Franaszek, 251-269. Warsaw: Jagellonian University Press, 2009
Brus, Wlodzimierz. "1950 to 1953: The Peak of Stalinism." In *The Economic History of Eastern Europe 1919-1975*. Vol. 3: *Institutional Change within a Planned Economy*, eds. Michael Charles Kaser and Edward A. Radice, 3-39. Oxford: Clarendon Press, 1985.
Bukharin, Nikolaĭ Ivanovic. "Notes of an Economist." In *Selected Writings on the State and on the Transition to Socialism*, ed. Richard B. Day, 301-309. New York: M.E. Sharpe Inc., 1982.

Burawoy Michael and János Lukács. *The Radiant Past: Ideology and Reality in Hungary's Road to Capitalism*. Chicago: University of Chicago Press, 1992.
Burawoy Michael. *The Politics of Production: Factory Regimes Under Capitalism and Socialism*. London: Verso, 1985.
Čapek, Karel. *RUR Rossum's Universal Robots*. Praha: Ot. Štorch-Marien, 1920.
Carr, Edward Hallett. *The Russian Revolution from Lenin to Stalin 1917-1929*, New York : Free Press, 1979.
Cataluccio, Francesco, and Francesca Gori. *La Primavera di Praga*. Milano: Franco Angeli, 1990.
Cedrych, Mario René, and Lukáš Nachtmann. *Škoda. Auta známá i neznámá. Prototypy i seriové automobily vyráběné od roku 1934*. Praha: Grada, 2003.
Chandler, Alfred Dupont, Jr.. *Scale and Scope : The Dynamics of Industrial Capitalism*. Cambridge: Belknap Press, 1994.
Chandler, Alfred Dupont, Jr.. *The Visible Hand: The Managerial Revolution in American Business*. Cambridge: Belknap Press, 1977.
Chandler, Alfred Dupont, Jr.. *Strategy and Structure: Chapters in the History of Industrial Enterprise*. Cambridge: Massachusetts Institute of Technology Press, 1962.
Chandler, Alfred Dupont, Jr. , Franco Amatori, and Takashi Hikino, *Big Business and the Wealth of Nations*. Cambridge: Cambridge University Press, 1997.
Cohen, Yves. "Administration, politique, et techniques: réflexions sur la materialité des pratiques administratives dans la Russie staliniénne (1922-1940)." *Cahiers du Monde Russe* 44, no. 2/3 (2003): 269-307.
Cohen, Yves. "Les Chefs, une question pour l'histoire du XXe siècle." *Cités* 6 (2001/2002): 67-83.
Cohen, Yves. "The Soviet Fordson: Between the Politics of Stalin and the Philosophy of Ford, 1924-1932." In *Ford of Europe 1903-2003: The European History*, edited by Hubert Bonin, Yannick Lung, and Steven Tolliday, 531-558. Paris: Plage, 2003.
Cohen, Yves. *Organiser à l'aube du Taylorisme. La pratique d'Ernest Mattern chez Peugeot 1906-1919*. Besancon: Annales Littéraires de l'Université de Franche-Compté, 2001.
Committee on Elimination of Waste in Industry of the Federated American Engineering Societies. *Waste in Industry*. New York: McGraw-Hill Book Company, 1921.
Crowley, David, and Susan Reid, eds. *Style and Socialism: Modernity and Material Culture in Post-War Eastern Europe*. Oxford: Berg, 2000.
Crowley, David, and Susan Reid, eds. *Socialist Spaces: Sites of Everyday Life in the Eastern Bloc*. Oxford: Berg, 2002.
Douglas, Dorothy. *Transitional Economic Systems: The Polish Czech Example*. London: Routledge, 1953.
Dubreuil, Hyacinthe. *L'Exemple de Bat'a. La Libération des initiatives individuelles dans une Entreprise Géante*. Paris: B. Grasset, 1936.
Edensor, Tim. "Automobility and National Identity: Representation, Geography and Driving Practice." *Theory, Culture and Society* 21, no. 4/5 (2004): 101-120.
Fava, Valentina. "Il dibattito tecnico sull'organizzazione del lavoro in Cecoslovacchia dal 1918 al 1948." Unpublished Tesi di Laurea, Università degli Studi di Milano, 1999.
Fava, Valentina. "The Automobile Industry and CMEA Integration. Evidence from the Czechoslovak Case and Reflections on an Unexpected Failure." *Jahrbuch für Wirtschaftsgeschichte / Economic History Yearbook* 2 (2008): 93-115.
Fejtö, François. *Histoire des démocraties populaires*. Paris, Editions du Seuil, 1952.
Fitzpatrick, Sheila. *The Russian Revolution*. Oxford: Oxford University Press, 1994.
Fleischner Jindřich, *Technická kultura. Sociálně-filosofické a kulturně-politické úvahy o dějinách technické práce*. Praha: Fr. Borový 1916.

Fléron, Frederick. *Technology and Communist Culture: The Socio-Cultural Impact of Technology under Socialism*. New York: Praeger, 1976.
Fligstein, Neil. *The Transformation of Corporate Control*. Cambridge, MA: Harvard University Press, 1990.
Flores, Marcello. *1956*. Bologna: Il Mulino, 1996.
Folta, Jaroslav, ed. *Studie o technice v Českých Zemích* Vol. 5, 457-463. Praha: Národní Technické Muzeum, 1918-1945.
Freeland, Robert F. "The Myth of the M-Form? Governance, Consent, and Organizational Change." *American Journal of Sociology* 102, no. 2 (September 1996): 483-526.
Freeland, Robert F. *The Struggle for Control of the Modern Corporation: Organizational Change at General Motors, 1924-1970*. Cambridge: Cambridge University Press, 2001.
Gaddis, John Lewis. *The United States and the Origins of the Cold War, 1941-1947*. New York: Columbia University Press, 1972.
Gilbreth, Frank Bunker and Lillian Gilbreth. *Motion Study*. New York: Hive, 1911.
Graham, Loren. *The Ghost of Executed Engineer: Technology and the Fall of the Soviet Union*. Cambridge, MA: Harvard University Press, 1993.
Gramsci, Antonio. *Americanismo e Fordismo*. Roma: Editori Riuniti, 1991.
Granick, David. *The Red Executive: A Study of the Organization Man in Russian Industry*, Garden City, N.Y., Doubleday, 1960.
Graziosi, Andrea. *Guerra e rivoluzione in Europa. 1905-1956*. Bologna: Il Mulino, 2001.
Graziosi, Andrea. *Stato e Industria in Unione Sovietica (1917-1953)*. Napoli: Edizioni Scientifiche Italiane, 1994.
Gross, Jan. "The Social Consequences of War: Preliminaries for the Study of the Imposition of Communist Regimes in East Central Europe." *East European Politics and Societies* 3 (1989): 198-214.
Haraszti, Miklos. *Worker in a Worker's State: Piecework in Hungary*, New York, Universe Books, 1975.
Hatzfeld, Nicolas. *Les gens d'usine. 50 ans d'histoire à Peugeot- Sochaux*. Paris: Les Editions de l'atelier, 2002.
Heumos, Peter. *„Vyhrňme si rukávy, než se kola zastaví!" Dělníci a státní socialismus v Československu, 1945-1968*. Praha: Ústav pro soudobé dějiny AV ČR, 2006.
Hogan, Michael J. *The Marshall Plan: America, Britain, and the Reconstruction of Europe*. Cambridge: Cambridge University Press, 1987.
Holý, Ladislav. *The Little Czech and the Great Czech Nation: National Identity and the Post-Communist Social Transformation*. Cambridge: Cambridge University Press, 1996.
Hounshell, David A. "Automation, Transfer Machinery, and Mass Production in the US Automobile Industry in the Post-World War II Era." *Enterprise and Society* 1, no. 1 (March, 2000): 100-138.
Hounshell, David A. "Planning and Executing 'Automation' at Ford Motor Company, 1945-1965: The Cleveland Engine Plant and Its Consequences." In *Fordism Transformed: The Development of Production Methods in the Automobile Industry*, edited by Haruhito Shiomi and Kazuo Wada, 49-86. Oxford: Oxford University Press, 1995.
Hounshell, David A. *From the American System to Mass Production: 1800-1932*. Baltimore: The John Hopkins University Press, 1985.
Jajesniak-Quast, Dagmara. "Permeating the Embargo with Iron and Steel: Poland, Czechoslovakia, the GDR and the Neutral States." In *Gaps in the Iron Curtain: Economic Relations between Neutral and Socialist Countries in Cold War Europe*, edited by Gertrude Enderle-Burcel, 270-289. Warsaw: Jagiellonian University Press, 2008.
Janáček, František, ed.. *Největší zbrojovka monarchie. Škodovka v dějinách, dějiny ve Škodovce. 1859-1918*. Plzeň: Nakladatelství Novinář, 1990.

Janko, Jan, and Emilie Těšínská, eds.. *Technokracie v Českých zemích (1900-1950)*. Praga: Archiv AV CR, 1999.
Jirásek, Jaroslav. *Organizační řízení. Vývoj teorie podnikového řízení*. Praha: Institut řizení, 1987.
Jíša, Václav. *Škodovy závody 1859-1865*. Praha: Práce 1965.
Josephson, Paul R. "Projects of the Century in Soviet History: Large Scales Technologies from Lenin to Gorbachev." *Technology and Culture* 34, no. 3 (1995): 519-559.
Kafka, Franz. Amerika. Munchen: K. Wolff, 1927.
Kaplan, Karel, and Pavel Paleček. *Komunistický režim a politické procesy v Československu*. Brno: Barrister - Principal, 2001.
Kaplan, Karel. "La crisi cecoslovacca." In *Annali della Fondazione Feltrinelli*, 267-326. Milano: Feltrinelli, 1982.
Kaplan, Karel. *Československo a RHVP (1948-1953)*. Praha: Ústav pro soudobé dějiny AV ČR, 1995.
Kaplan, Karel. *Československo v poválečné Evropě*. Praha: Karolinum, 2004.
Kaplan, Karel. *Pět kapitol o únoru*. Praha: Ústav pro soudobé dějiny, 1997.
Kaplan, Karel. *Pravda o Československo, 1945-1948*. Praha: Panorama, 1992.
Kaplan, Karel. *Sociální souvislosti krizí komunistického režimu v letech 1953-1957 a 1968-1975*. Praha: Ústav pro soudobé dějiny AV ČR, 1993.
Kaplan, Karel. *Sovětšti poradci v Československu. 1949-1956*. Praha: Ústav pro soudobé dějiny AV ČR, 1993.
Kaplan, Karel. *The Overcoming of the Regime Crisis after Stalin's Death in Czechoslovakia, Poland and Hungary*. Praha: Panorama, 1992.
Kaplan, Karel. *The Short March: The Communist Takeover in Czechoslovakia 1945-1948*. London: C. Hurst, 1987.
Karlický, Vladimír, Petr Hofman, František Janáček, Antonín Klimek, and Vladislav Krátký. *Svět okřídleného šípu. Koncern Škoda Plzeň. 1918-1945*. Plzeň: Škoda a. s. ve spolupráci s nakladatelstvím Paseka, 1999.
Kárník, Zdeněk, and J. Měchýř. *K novověkým sociálním dějinám Českých zemí IV. Zvraty a převraty 1939-1992*. Praha: Karolinum, 2001.
Kárník, Zdeněk, and Michal Kopeček. *Bolševismus, komunismus a radikální socialismus v Československu*, Vol. 1. Praha: Ústav pro soudobé dějiny AV ČR, Dokořán, 2003.
Kaser, Michael Charles. *Comecon. Problemi di integrazione nelle economie pianificate*. Milano: Franco Angeli, 1976.
Kipping, Matthias, and Ove Bjarnar, eds. *The Americanisation of European Business: The Marshall Plan and the Transfer of US Management Models*. London: Routledge, 1998.
Kirschbaum, Stanislav J., ed. *Central European History and the European Union: The Meaning of Europe*. Basingstoke: Palgrave, 2007.
Kornai, János. *The Socialist System: The Political Economy of Communism*. Princeton: Princeton University Press, 1992.
Kosta, Jiří. "Storia e contenuti della riforma economica cecoslovacca negli anni 1965-1969." In *Che cosa fu la "Primavera di Praga,"* edited by Francesco Leoncini, 41-67. Manduria: Lacaita, 1989.
Kotkin, Stephen. *Magnetic Mountain: Stalinism as a Civilization*. Berkeley: University of California Press, 1995.
Kott, Sandrine. "Pour une histoire sociale du pouvoir en Europe Communiste: introduction thématique. " *Revue d'Histoire Moderne et Contemporaine* 2 (2002): 5-23.
Kott, Sandrine. *Le communisme au quotidien. Les entreprises d'Etat dans la societé est-allemande*. Paris: Belin, 2001.
Kovanda, Karel. "I consigli operai in Cecoslovacchia (1968-1969)." In *Che cosa fu la*

"Primavera di Praga," edited by Francesco Leoncini, 69-90. Manduria: Lacaita, 1989.
Kožíšek, Petr, and Jan Králik. *L&K-Škoda,* vol. 1 and 2. Prague: Motor Press, 1997.
Králík, Jan. *Utajené projekty Škoda. Strhující příběh konstruktéra Oldřicha Meduny.* Praha: Grada Publishing, 2007.
Králík, Jan. *V soukolí okřídleného šípu. Zákulisí automobilky Škoda očima konstruktéra Favoritu Petra Hrdličky.* Praha: Grada Publishing, 2008.
Kramer, Mark. "New Sources on the 1968 Soviet Invasion of Czechoslovakia (Part One)." In *Cold War International History Project Bulletin* 2 (Fall 1992). Washington, DC: Woodrow Wilson International Center for Scholars, 1992.
Kramer, Mark. "New Sources on the 1968 Soviet Invasion of Czechoslovakia (Part Two)." In *Cold War International History Project Bulletin* 2 (Fall 1993). Washington, DC: Woodrow Wilson International Center for Scholars, 1992.
Kramer, Mark. "The Early Post-Stalin Succession Struggle and Upheavals in East Central Europe: Internal-External Linkages in Soviet Policy Making." *Journal of Cold War Studies* 1, no. 1 (Winter, 1999): 3-55.
Kramer, Mark. "The Early Post-Stalin Succession Struggle." *Journal of Cold War Studies* 1, no. 3 (Fall, 1999): 3-66.
Krejčí, Jaroslav. *Social Change and Stratification in Postwar Czechoslovakia.* London: Macmillan, 1972.
Kubů, Eduard, and Jaroslav Pátek. *Mýtus a realita hospodářské vyspělosti Československa mezi světovými válkámi.* Praha: Karolinum, 2000.
Lacina, Vlastislav. "The Impact of the World Economic Crisis of the 1930s on the Economy of Czechoslovakia." *Hospodářske Dějiny-Economic History* 15 (1986): 145-172.
Lacina, Vlastislav. "The Tempo of Economic Growth in Inter-War Czechoslovakia: National Income and Gross Domestic Product as Basic Measures of General Economic Development." *Historica* 2, no. 32 (1995): 195-222.
Lacina, Vlastislav. *Formování československé ekonomiky, 1918-1923.* Praha: Historický ústav AV ČR, 1994.
Lampert, Nicholas. *The Technical Intelligentsia and the Soviet State: A Study of Soviet Managers and Technicians, 1928-1935.* London: Macmillan, 1979.
Laudiero, Alfredo. *Api e cinghiali. Culture politiche dell'Europa centrale: il caso ceco.* Soveria Mannelli: Rubettino, 2000.
Laux, James Michael. *The European Automobile Industry.* New York: Twayne Publishers, 1992.
Lenin, Vladimir I. "I compiti immediati del potere sovietico," *Opere scelte.* Moscow: Progress Publishers, 1971, 412-443.
Lenin, Vladimir I., "Stato e rivoluzione. La dottrina marxista dello stato e i compiti del proletariato nella rivoluzione." *Opere scelte.* Moscow: Progress Publishers, 1971, 268-359.
Lenin, Vladimir I. "What is Soviet Power?" In *Collected Works,* 4th English Edition, Volume 29, 248-249. Moscow: Progress Publishers, 1972. (Speech, end of March, 1919.)
Lenin, Vladimir I. *Collected Works,* 4th English Edition. Moscow: Progress Publishers, 1972.
Leoncini, Francesco, ed. *Alexander Dubček e Jan Palach protagonisti della storia europea.* Soveria Mannelli (Cz): Rubbettino, 2009.
Leoncini, Francesco, ed. *Che cosa fu la "Primavera di Praga." Idee e progetti di una riforma politica e sociale.* Manduria: Lacaita, 1989.
Leoncini, Francesco, Luciano Antonetti, and Carla Tonini. *Primavera di Praga e dintorni. Alle origini dell'89.* S. Domenico di Fiesole: Edizioni Cultura della Pace, 2000.
Leoncini, Francesco. "1918-1968. Le due Primavere della Nuova Europa." In *Primavera di Praga e dintorni. Alle origini dell'89,* edited by Francesco Leoncini, Luciano Antonetti, and Carla Tonini, 13-35. S. Domenico di Fiesole: Edizioni Cultura della Pace, 2000.
Leoncini, Francesco. "Federalism in Central Europe: Past and Present." In *Central European*

History and the European Union: The Meaning of Europe, edited by Stanislav J. Kirschbaum. Basingstoke: Palgrave, 2007.

Leoncini, Francesco. "Presentazione al saggio di Alberto Bazala, Il Pensiero di Masaryk, pubblicato a Praga nel 1935." *Slavia* 14 (2005): 104-151.

Leoncini, Francesco. "T.G. Masaryk's Nová Evropa: A Reinterpretation." In *Historical Reflections on Central Europe: Selected Papers from the Fifth World Congress of Central and East European Studies, Warsaw, 1995*, edited by Stanislav J.Kirschbaum, 65-73. Basingstoke: Macmillan, 1999.

Leoncini, Francesco. *La questione dei Sudeti, 1918-1938*. Padova: Liviana, 1976.

Leoncini, Francesco. *L'Europa Centrale. Conflittualità e progetto. Passato e presente tra Praga, Budapest e Varsavia*. Venezia: Cafoscarina, 2003.

Lewin, Moshe.. *The Making of the Soviet System: Essays in the Social History of Interwar Russia* (London: Methuen, 1985).

Locke, Robert R. *The Collapse of the American Management Mystique*. Oxford: Oxford University Press, 1996.

Lukeš, Igor. *Czechoslovakia between Stalin and Hitler: The Diplomacy of Edvard Beneš in the 1930s*. Oxford: Oxford University Press, 1996.

Machat, František. *Dějiny vědeckého řízení v kapitalistickém průmyslu – Vybrané kapitoly*. Praha: Nakladatelství Svoboda, 1967.

Machonin, Pavel, and Jaroslav Krejčí. *Czechoslovakia 1918-1992: A Laboratory for Social Change*. Basingstoke: Macmillan, 1996.

Machonin, Pavel. *Československá společnost: sociologická analýza sociální stratifikace*. Bratislava: Epocha, 1969.

Maier, Charles S. "Between Taylorism and Technocracy: European Ideologies and the Vision of Industrial Productivity in the 1920s." *Journal of Contemporary History* 5, no. 2 (1970): 27-61.

Maier, Charles S. *In Search of Stability: Explorations in Historical Political Economy*. Cambridge: Cambridge University Press, 1987.

Maier, Charles. „The Politics of Productivity: Foundation of American International Policy after World War II." *International Organization* 31 (September 1977): 607-633.

Mamatey, Victor S., and Radomír Luža, eds. *A History of the Czechoslovak Republic, 1918-1948*. Princeton: Princeton University Press, 1973.

Maňák, Jiří. "Proměna dělnické strany v organizzaci moci. Problematika dělnického charakteru KSČ v letech 1945-1953 ve světle stranické statistiky." In *Bolševismus, komunismus a radikální socialismus v Československu*, edited by Zdeněk Kárník and Michal Kopeček, 157-199. Praha: Ústav pro soudobé dějiny AV ČR, Dokořán, 2003.

Margolius, Ivan, and Charles Meisl. *Škoda-Laurin and Klement*. London: Osprey Automotive, 1992.

Martelli, Antonio. *Teorie e ideologie del management. Profilo storico delle dottrine manageriali (1770-1970)*. Milano: ETAS libri, 1979.

Masaryk, Tomáš Garrigue. *Česká otázka*. Praha: Svoboda, 1990.

Masaryk, Tomáš Garrigue. *La nuova Europa: il punto di visto slavo*, edited by F. Leoncini. Pordenone: Edizioni Studio Tesi, 1997.

Masaryk, Tomáš Garrigue. *Problém malého národa*. Praha: Neutralita, 1990.

Mayer, Françoise. *Les Tchèques et leur Communisme. Mémoire et identités politiques*. Paris: Editions de l'Ecole des Hautes Etudes en Sciences Sociales, 2004.

McDermott, Gerald A. "Renegotiating the Ties That Bind: The Limits to Privatization in the Czech Republic." In *Restructuring Networks in Post-Socialism: Legacies, Linkages, and Localities*, edited by Gernot Grabher and David Stark, 77. Oxford: Oxford University Press, 1997.

McDermott, Gerald A. *Embedded Politics: Industrial Networks and Institutional Change in Postcommunism*. Ann Arbor: Michigan University Press, 2002.
McDermott, Gerald. "Network Restructuring and Firm Creation in East Central Europe: A Public Private Venture." In *The New Entrepreneurs of Europe and Asia*, edited by Victoria Bonnell and Thomas B.Gold, 214-255. New York: M.E.Scarpe, 2002.
McGlade, Jacqueline. "Americanization: Ideology or Process? The Case of the United States Technical Assistance and Productivity Programme." In *Americanization and its Limits*, edited by Jonathan Zeitlin and Gary Herrigel, 53-75. Oxford: Oxford University Press, 2000.
McGlade, Jacqueline. "Lo zio Sam ingegnere industriale. Il programma americano per la produttività e la ripresa economica dell'Europa occidentale (1948-1958)." *Studi Storici* 37, no. 1 (January-March 1996): 9-40.
McGlade, Jacqueline. "The US Technical Assistance Program: From Revolutionary Vision to Production Drive." In *Catching up with America: Productivity Missions and the Diffusion of American Economic and Technological Influence after the Second World War*, edited by Dominique Barjot, 67-86. Paris: Presses de l'Université de Paris-Sorbonne, 2002.
Meer, Elisabeth Van. "Engineering Beyond Politics: Professional Ideology, Scientific Management and the Evolution of Czechoslovakia." Unpublished PhD diss., University of Minnesota, 2006.
Merkel, Ida. "Consumer Culture in the GDR, or How the Struggle for Antimodernity was Lost on the Battleground of Consumer Culture." In *Getting and Spending: European and American Consumer Societies in the Twentieth Century*, edited by Susan Strasser, Charles McGovern, and Matthias Judt, 281-301. Washington DC: Publications of the German Historical Institute, 1998.
Merkle, Judith A. *Management and Ideology: The Legacy of the International Scientific Management Movement*. Berkeley: University of California Press, 1980.
Michal, Jan M. "Postwar economic development." In *A History of the Czechoslovak Republic, 1918-1948*, edited by Victor S. Mamatey and Radomir Luža, 428-460. Princeton: Princeton University Press, 1973.
Milward, Alan S. *The Reconstruction of Western Europe, 1945-1951*. London: Routledge, 1984.
Myant, Martin R. *The Czechoslovak Economy 1948-1988. The Battle for Economic Reform*. Cambridge: Cambridge University Press, 1989.
Narinski, Mikhail. "Le Plan Marshall et l'URSS." In *Le Plan Marshall et le relèvement économique de l'Europe. Colloque tenu à Bercy les 21, 22, et 23 mars 1991*, edited by René Girault and Maurice Lévy-Leboyer, 119-123. Paris: Comité pour l'histoire économique, 1993.
Navrátil, Jaromír, ed. *The Prague Spring, 1968: A National Security Archive Documents Reader*. Budapest: Central European University Press, 1998.
Nee, Victor, and David Stark, eds. *Remaking the Economic Institutions of Socialism, China and Eastern Europe*. Stanford: Stanford University Press, 1989.
Nelson, Daniel. *Frederick W. Taylor and the Rise of Scientific Management*. Madison: University of Wisconsin Press, 1980.
Nolan, Mary. *Visions of Modernity: American Business and the Modernization of Germany*. Oxford: Oxford University Press, 1994.
Ojala, Jari, and Jari Eloranta. *East West Trade and the Cold War*. Jyväskylä: University of Jyväskylä, 2005.
Oldenziel, Ruth, and Karin Zachmann, eds. *Cold War Kitchen: Americanization, Technology and European Users*. Cambridge, MA: The MIT Press, 2009.
Olmo, Carlo, ed. *Mirafiori*. Torino: Allemandi, 1997.
Overy, Richard. "German Multinationals and the Nazi State in Occupied Europe." In *Multinational Enterprise in Historical Perspective*, edited by Alice Teichová, Maurice Lévy-Leboyer, and Helga Nussbaum. Cambridge: Cambridge University Press, 1989.

Paďourek, Jan. "International Contacts of the Czech Technical Academy: The Masaryk Labour Academy and the World in between the Two World Wars." *Studiae Historiae Academiae Scientiarum Bohemicae* 2 (1993): 35-50.
Paďourek, Jan. "Snahy o organizování praxe československých inženýrů v USA (20 a 30 léta 20. století)." *Dějiny věd a techniky* 3 (1992): 129-139.
Pavitt, Keith. "Alcuni fondamenti della teoria della grande impresa innovativa." In *Innovazione, impresa e sviluppo economico*, edited by Renato Giannetti and Pierangelo Maria Toninelli. Bologna: Il Mulino, 1991.
Pernes, Jiři. "Dělnické demonstace v Brně v roce 1951." *Soudobé dějiny* 1 (1996): 23-41.
Pernes, Jiři. *Snahy o překonaní politicko-hospodářské krize v Československu v roce 1953*. Brno: Ústav pro soudobé dějiny, AV ČR, 2000.
Pittaway, Mark "Introduction: Workers and Socialist States in Postwar Central and Eastern Europe." *International Labor and Working Class History* 68 (2005): 1-8.
Pittaway, Mark "Workers, Management and the State in Socialist Hungary: Shaping and Re-shaping the Socialist Factory Regime in Újpest and Tatabánya, 1950-1956" in Christiane Brenner and Peter Heumos eds. *Sozialgeschichtliche Kommunismusforschung: Tschechoslowakei, Polen, Ungarn und DDR 1945-1968*, 105-131. München: R. Oldenbourg Verlag, 2005.
Pollard, Sidney. *Peaceful Conquest*. Oxford: Oxford University Press, 1981.
Price, Brian. "Frank and Lillian Gilbreth and the Motion Study Controversy, 1907-1930." In *A Mental Revolution: Scientific Management since Taylor*, edited by Daniel Nelson, 58-76. Columbus: Ohio State University Press, 1992.
Prucha, Václav. *Hospodářské dějiny Československa v 19. a 20. Storoči*. Bratislava: Svoboda, 1974.
Průcha, Václav. *Sectoral Changes in Industry after World War II*. Prague: Institute of History of the ČSAS, 1991.
Radomír, Luža. "Czechoslovakia between Democracy and Communism, 1945-1948." In *A History of the Czechoslovak Republic, 1918-1948*, edited by Victor S. Mamatey and Radomir Luža, 387-415. Princeton: Princeton University Press, 1973.
Rathenau, Walther. *Die neue Wirtschaft*, Berlin: S. Fischer, 1918.
Roeser, Matthias. "Sovětizace a militarizace československého leteckého průmyslu v letech 1949-1953." *Soudobé Dějiny* 1 (1996): 60-71.
Rosenberg, Nathan. *Inside the Black Box: Technology and Economics*. Cambridge: Cambridge University Press, 1982.
Rosický, Miroslav. "Organization and Management of Industry and Construction in the Czechoslovak Socialist Republic." *Czechoslovak Economic Papers* 2 (1962): 60-61.
Rosický, Miroslav. *Organizace výrobní základny a plánovité řízení ekonomiky*. Praha: Akademie, 1983.
Rupnik, Jacques. "The Roots of Czech Stalinism." In *Essays for Eric Hobsbawm: Culture, Ideology, and Politics*, edited by Raphael Samuel and Gareth Stedman Jones, 304 and 310-311. London: Routledge, 1982.
Rupnik, Jacques. *Histoire du Parti communiste Tchecoslovaque: des origines à la prise du pouvoir*. Paris: Presses de la Fondation nationale des sciences politiques, 1981.
Salsano, Alfredo. *Ingegneri e politici. Dalla razionalizzazione alla "rivoluzione manageriale."* Torino: Einaudi, 1987.
Sapelli, Giulio. *Organizzazione, lavoro e innovazione industriale nell'Italia tra le due guerre*. Torino: Rosenberg & Sellier, 1978.
Schmidt, Eva. *Thomas G. Masaryk's Realism: Origins of a Czech Political Concept*. München: Oldenbourg Verlag, 1984.
Schot, Johan, and Vincent Lagendijk. "Technocratic Internationalism in the Interwar Years:

Building Europe on Motorways and Electricity Networks." *Journal of Modern European History* 2 (2008): 196-216.
Schumpeter, Joseph Alois. *Capitalismo, socialismo e democrazia*. Milano: Etas, 2001.
Schwartz, Michael. "Markets, Networks, and the Rise of Chrysler in Old Detroit, 1920-1940." *Enterprise and Society* 1 (March, 2000): 63-99.
Scranton, Phil. *Endless Novelty: Specialty Production and American Industrialization, 1865-1925*. Princeton: Princeton University Press, 1997.
Shiomi, Haruhito and Kazuo Wada, eds. *Fordism Transformed: The Development of Production Methods in the Automobile Industry*. Oxford: Oxford University Press, 1995.
Shpotov, Boris. "Ford in Russia from 1909 to World War II." In *Ford of Europe 1903-2003: The European History*, edited by Hubert Bonin, Yannick Lung, and Steven Tolliday, 525-529. Paris: Plage, 2003.
Siegelbaum, Lewis H. "Master of the Shop Floor: Foremen and Soviet Industrialization." In *Stalinism: Its Nature and Aftermath. Essays in honour of Moshe Lewin*, edited by Nicholas Lampert and Gábor Tamás Rittersporn, 127-156. London: Macmillan, 1992.
Siegelbaum, Lewis H. *Cars for Comrades: The Life of the Soviet Automobile*. Ithaca: Cornell University Press, 2008.
Siegelbaum, Lewis H. *Stakhanovism and the Politics of Productivity in the USSR, 1935-1941*. Cambridge: Cambridge University Press, 1988.
Šik, Ota. *Economic Planning and Management in Czechoslovakia*, Prague: Orbis, 1966.
Šik, Ota. *Plán a trh za socializmu*. Praha: Akademia, 1968.
Šik, Ota. *Quale comunismo?* Bari: Laterza, 1977.
Skilling, Gordon. *Czechoslovakia, 1918-1988: Seventy Years from Independence*. New York: St. Martin's Press, 1991.
Skilling, Gordon. *Czechoslovakia's Interrupted Revolution*. Princeton: Princeton University Press, 1976.
Smrček, Otto. "Albín Bašus, propagator vědecké organizace práce." *Dějiny věd a techniky* 3 (1992): 170-177.
Smrček, Otto. "Labor-ethics: A Czechoslovak Analogy of Technocracy." *Acta Historiae rerum naturalium necnon technicarum* 21 (1989): 145-161.
Smrček, Otto. "Vědecká organizace práce a její aplikace ve strojírenství do konce druhé světové války." *Hospodarské dějiny-Economic History* 13 (1985): 165-219.
Smrček, Otto. *Kapitoly z dějin strojírenství*. Prague: Agentura RETIP, 1992.
Špaček, Stanislav, ed. *Ford a My. Zkušenosti československých inženýrů z americké prakse u Forda s ohledem na naše poměry*. Praha: Jednota přátel Masarykovy akademie práce, 1928.
Špaček, Stanislav, ed. *Život a práce u Forda. Zápisky československých inženýrů z Ameriky*. Praha: Jednota přátel Masarykovy akademie práce, 1927.
Špaček, Stanislav. *Masarykovà Akademie Práce, knihovny Práce*, Vol. 1, 7-8. Praha: Nákl. Josefa Mrkvičky, 1919.
Špaček, Stanislav. *První ročenka masarykovy akademie práce. Zpráva o založení a činnosti za léta 1920 až 1923*. Praha: 1924.
Stalin, Josip V., *Questioni del Leninismo*. Ed.Unità: Roma, 1945.
Státník, Dalibor. *Sankční pracovní právo v padesátých letech. Vládní nařízení o opatřeních proti fluktuaci a absenci č.52\1953 Sb*. Praha: Ústav pro soudobé dějiny AV ČR, 1999.
Steiner, André. "Dissolution of the ‚Dictatorship over Needs?" Consumer Behavior and Economic Reform in East Germany in the 1960s." In *Getting and Spending: European and American Consumer Societies in the Twentieth Century*, edited by Susan Strasser, Charles McGovern, and Matthias Judt, 167-185. Washington DC: Publications of the German Historical Institute, 1998.

Stokes, Raymond. "In Search of the Socialist Artifact: Technology and Ideology in East Germany, 1945-1962." *German History* 15 (1997): 221-239.
Stokes, Raymond. "Plastics and the New Society: The German Democratic Republic in the 1950s and the 1960s." In *Style and Socialism: Modernity and Material Culture in Post-War Eastern Europe*, edited by David Crowley and Susan Reid, 65-80. Oxford: Berg, 2000.
Stokes, Raymond. *Constructing Socialism: Technology and Change in East Germany 1945-1990*. Baltimore: The Johns Hopkins University Press, 2000.
Strasser, Susan, Charles McGovern, and Matthias Judt, eds. *Getting and Spending: European and American Consumer Societies in the Twentieth Century*. Washington DC: Publications of the German Historical Institute, 1998.
Tchalakov Ivan, Tihomir Mitev and Venelin Petrov. "The Academic Spin-Offs as an Engine of Economic Transition in Eastern Europe. A Path-Dependent Approach." *Minerva*, 48, 2 (2010): 189-217.
Tchalakov, Ivan. "Bulgaria." In *"Mittelstand" and Trade in South Eastern Europe: Social Market Economy as a Compass of Development*, edited by Anna Katharina Bolling, 47-62. Berlin: Conrad Adenauer Stiftung, 2007.
Tchalakov, Ivan. "Technology and the Models of Post-Socialist Economic Transition: Some Evidences from the Bulgarian experience." in *Governing Sociotechnical Change in Southeastern Europe: Contributions from a Science & Technology Studies Perspective*, edited by Ivan Tchalakov, Harald Rohracher, and Franc Mali 169-201. Sofia: East-West Publishers, 2007.
Tchalakov Ivan. "Socialism as Society of Networks and the Problem of Technological Innovations." *Sociological Problems* (2006): 343-371.
Tchalakov, Ivan "The Socialist Entrepreneurs Viewed From the Joint Perspective of Joseph Schumpeter, Actor-Network Theory and Historical Sociology of Socialism." Research in *Social Change*, 3, 1 (2011): 15-51.
Teichová, Alice, Maurice Lévy-Leboyer, and Helga Nussbaum, eds. *Multinational Enterprise in Historical Perspective*. Cambridge: Cambridge University Press, 1986.
Teichová, Alice, and Herbert Matis, eds. *Österreich und die Tschechoslowakei: 1918-1938. Die wirtschaftliche Neuordnung in Zentraleuropa in der Zwischenkriegszeit*. Wien: Böhlau Verlag, 1996.
Teichová, Alice. "Les hésitations tchèques sur le Plan Marshall." In *Le Plan Marshall et le relèvement économique de l'Europe. Colloque tenu à Bercy les 21, 22, et 23 mars 1991*, edited by René Girault and Maurice Lévy-Leboyer, 109-118. Paris: Comité pour l'histoire économique, 1993.
Teichová, Alice. "The Halting Pace to Scope and Scale." In *Big Business and the Wealth of Nations*, edited by Alfred Dupont Chandler, Franco Amatori, and Takashi Hikino, 447-461. Cambridge: Cambridge University Press, 1997.
Teichová, Alice. *An Economic Background to Münich: International Business and Czechoslovakia 1918-1938*. Cambridge: Cambridge University Press, 1974.
Teichová, Alice. *Německá hospodářská politika v Českých zemích v letech 1939-1945*. Praga: VSE, 1998.
Teichová, Alice. *The Czechoslovak Economy 1918-1980*. London: Routledge, 1984.
Teichová, Alice, Gertrude Enderle-Burcel, and Piotr Franaszek, *Gaps in the Iron Curtain: Economic Relations between Neutral and Socialist Countries in Cold War Europe*. Warsaw: Jagellonian University Press, 2009.
Těšínská, Emilie. "Vznik Masarykovy akademie práce: technokratické tendence a účast přírodovědců." In *Technokracie v Českých zemích (1900-1950)*, edited by Jan Janko and Emilie Těšínská, 108-110. Praha: Archiv AV ČR, 1999.
Tolliday, Steven, ed. *The Rise and Fall of Mass Production*. Cheltenam: Edward Elgar, 1998.

Toninelli, Pier Angelo, ed. *The Rise and Fall of State-Owned Enterprise in the Western World.* Cambridge: Cambridge University Press, 2000.
Triska, Jan F., and Charles Gati, eds. *Blue-Collar Workers in Eastern Europe.* London: George Allen and Unwin, 1981.
Ulč, Otto. "Plzeň: The Unknown Revolt." *Problems of Communism* 14, no. 3 (May-June, 1965): 46-49.
Urry, John. "The 'System' of Automobility." *Theory, Culture and Society* 21, no. 4/5 (2004): 25-39.
Verunáč, Václav. *Racionalisace, vědecká organisace a otázka sociální.* Prague: Sociální ústav republiky Českosl., 1930.
Verunáč, Václav. Ústav pro technické hospodářství průmyslové. Prague: Nákladem vlastním, 1926.
Veselý, Zdeněk. Československo a Marshallův Plán. Praga: Acta Universitatae Carolinae Philosophica et Historica Monographia XCII, 1982.
Viani, Paolo. "Progettare l'impresa: Francesco Mauro e il dibattito europeo tra le due guerre." In *Storie d'imprenditori*, edited by Duccio Bigazzi, 235-292. Bologna: Mulino, 1996.
Vigner, Miroslav "Vliv účelného uspořádání výrobního pochodu na hospodárnost." *Podniková organisace,* no. 7 (1955): 264-267.
Weber, Max. *Der Sozialismus. Rede zur allgemeinen Orientierung von oesterreichischen Offizieren in Wien 1918*, Gesammelte Aufsaete zur Soziologie und Sozialpolitik, 1924)
Werth, Nicolas. *Histoire de l'Union Soviétique: de l'Empire Russe* à la communauté des Etats indépendants. *1900-1991.* Paris: Presses Universitaires de France, 1990.
Whittington, Richard, and Michael Mayer. *The European Corporation: Strategy, Structure, and Social Science.* Oxford: Oxford University Press, 2000.
Woodward, Joan. *Industrial Organization: Theory and Practice.* Oxford: Oxford University Press, 1980.
Zatlin, Jonathan R. „The Vehicle of Desire: The Trabant, the Wartburg, and the End of the GDR." *German History* 15 (1997): 358-380.
Zeitlin, Jonathan, and Gary Herrigel, eds. *Americanization and its Limits: Reworking US Technology and Management in Post-War Europe and Japan.* Oxford: Oxford University Press, 2000.
Zeitlin, Jonathan. "Introduction." In *Americanization and its Limits*, 5. Oxford: Oxford University Press, 2000.
Zeitlin, Jonathan. "Reconciling Automation and Flexibility? Technology and Production in the Postwar British Motor Vehicle Industry." In *Enterprise and Society* 1 (March, 2000): 9-62.
Zunz, Olivier. *Perché il secolo americano?* Edited by Arnaldo Bagnasco. Translated by Di Giuseppina Cavallo. Bologna: Il Mulino, 2000. (*Why the American Century?* Chicago: The University of Chicago Press, 2000.)
Žalud, František H. *Přežili jsme.* Praha: Trilabit, 1998.

Endnotes

Chapter 1

1 "High volume manufacture of standardized goods using special purpose machinery and predominantly unskilled labour – together with the host of systematic management techniques, organizational structures and research and marketing services developed for its efficient administration and effective exploitation [...]." Jonathan Zeitlin and Gary Herrigel, eds., *Americanization and Its Limits: Reworking US Technology and Management in Post-War Europe and Japan* (Oxford: Oxford University Press, 2000), 2-3; Steven Tolliday, ed., *The Rise and Fall of Mass Production* (Cheltenam: Edward Elgar, 1998).

2 Petr Kožíšek and Jan Králík, *L&K-Škoda,* vol. 1 and 2 (Prague: Motor Press, 1997); Mario René Cedrych and Lukáš Nachtmann, *Škoda. Auta známá i neznámá. Prototypy i seriové automobily vyráběné od roku 1934* (Prague: Grada, 2003); Vladimír Karlický et al., *Svět okřídleného šípu. Koncern Škoda Plzeň. 1918-1945* (Plzeň: Paseka, 1999), 397.

3 "Výroba osobních a dopravních automobilů Škoda v Mladé Boleslavi," *Zprávy Československého národního komitétu pro vědeckou organisaci* 2 (1931): 17-18.

4 Škoda-Auto Historical Archive, Mladá Boleslav (henceforth Aša), Akciová společnost pro automobilový průmysl (henceforth ASAP), b. 24, see also: *Škodovák,* June 6, 1930, and "Pryč s racionalisací, prováděnou na účet dělnictva!" *Právo Lidu,* June 18, 1930.

5 Eduard Kubů and Jaroslav Pátek, *Mýtus a realita hospodářské vyspělosti Československa mezi světovými válkami* (Prague: Karolinum, 2000); Alice Teichová, *The Czechoslovak Economy 1918-1980* (London-New York: Routledge, 1984); Vlastislav Lacina, *Formováni československé ekonomiky, 1918-1923* (Prague: Historický ústav AV ČR, 1990); Vlastislav Lacina, "The Tempo of Economic Growth in Inter-War Czechoslovakia: National Income and Gross Domestic Product as Basic Measures of General Economic Development," *Historica* (series nova), no. 2\XXXII (1995): 214.

6 Anna Anfossi, *Prospettive sociologiche sull'organizzazione aziendale* (Milano: Franco Angeli, 1971), 42; Joan Woodward, *Industrial Organization: Theory and Practice* (Oxford: Oxford University Press, 1982); Franco Amatori, *Cicli produttivi, tecnologie, organizzazione del lavoro. La siderurgia a ciclo integrale dal piano autrchico alla fondazione dell'Italsider 1937-1961* (Firenze: Cooperativa editrice umanitaria, 1980), 564.

7 Francesco Leoncini, "Introduzione", in Tomáš Garrigue Masaryk, *La Nuova Europa,* trans. Franco Giuliano and Francesco Leoncini (Pordenone: Edizioni Studio Tesi, 1997), 37; see also, by the same author: *L'Europa centrale. Conflittualità e progetto. Passato e presente tra Praga, Budapest e Varsavia* (Venice: Cafoscarina, 2003); and especially the study *T.G. Masaryk precursore della nuova Europa,* 235-55.

8 Sidney Pollard, *La conquista pacifica. L'industrializzazione in Europa dal 1760 al 1970* (Bologna: Il Mulino, 1989), 316-17 (or. *Peaceful Conquest*, Oxford: Oxford University Press, 1981); Ivan Jakubec and Jaroslav Pátek, "Průmysl," in Kubů, Pátek, *Mýtus a realita hospodářské vyspělosti Československa mezi světovými válkami,* 9-134.

9 Teichovà, *The Czechoslovak Economy 1918-1980,* 21.

10 Lacina, *The Tempo of Economic Growth,* 214.

11 Otto Smrček, "Vědecká organizace práce a jeji aplikace ve strojírenstvi do konce druhé světové války," *Hospodářské dějiny* (Economic History) 13 (1985): 165-219; Otto Smrček, *Kapitoly z dějin strojírenství* (Prague: Agentura Retip, 1992), 2-31; František Machat, *Dějiny vědeckého řízení v kapitalistickém průmyslu – Vybrané kapitoly* (Prague: Nakladatelství Svoboda, 1967).

12 Albín Bašus, *Moderní organisace v strojírnách: studie technicko-hospodářská* (Prague: Nákladem vlastním, 1911); Jindřich Fleischner, "Technická kultura," *Technický ozbor* (1911): in the same year Fleischner founded the journal entitled: *Technika XX. Století: Revue pro kulturu, a technickou ekonomii.* See also for a more detailed perspective on the pre-war technical debate: Smrček, *Vědecká organizace práce,* 199, and Albín Bašus, propágator vědecké organizace práce," *Dějiny věd a techniky* 3 (1992): 170-77; Elisabeth Van Meer, "Engineering Beyond Politics: Professional Ideology, Scientific Management and the Evolution of Czechoslovakia," PhD thesis, University of Minnesota, 2006; Valentina Fava, "Il dibattito tecnico sull'organizzazione del lavoro in Cecoslovacchia dal 1918 al 1948," Tesi di Laurea, Università degli Studi di Milano, 1999.

13 Stanislav Špaček (1876-1954) was the most active among the Czechoslovak engineers who were interested in the studies and spread of scientific management in Czechoslovakia: Stanislav Špaček, "O pracovní a lidské ekonomii," *Cement, železo a beton*, no. 8 (1914); "O lidské ekonomii," *Vzdělání lidu*, no. 11 (1914); "Taylorismus a jeho význam pro práci v době mírové," *Obzor průmyslový*, no. 1 (1915-16); "Taylorismus ze stanoviska pracovní a lidské ekonomie," *Obzor průmyslový*, no. 1 (1915-16).

14 Emilie Těšínská, "Vznik Masarykovy akademie práce: technokratické tendence a účast přírodovědců," in *Technokracie v Českých zemích (1900-1950)*, ed. Jan Janko and Emilie Těšínská (Archiv AV ČR, Prague, 1999), 108-10; Stanislav Špaček, *Masarykovà akademie práce, knihovny Práce*, vol. 1 (Prague, 1919), 7-8; *První ročenka Masarykovy akademie práce. Zpráva o založení a činnosti za léta 1920 až 1923* (Prague, 1924). The Academy had a very complex structure. As a central governance body, it had a president, named by the President of the Czechoslovak Republic, a Scientific Council and a Body of Experts. Its activity was divided among many institutes, branches and commissions – originally six: 1) Natural and pharmaceutical sciences – *Ústav přírodovědecký a lékařský*; 2) Agriculture and forestry – *Ústav zemědělský a lesnický*; 3) Architecture and Engineering – *Ústav stavebně-inženýrský*; 4) Machines and electrotechnics – *Ústav strojní a elektrotechnický*; 5) Chemistry and technology – *Ústav chemicko-technologický*; 6) Institute for national and social economy – *Ústav národohospodářský a sociální.* But their number increased significantly during the twenty years of its existence. In terms of funding, originally the club of Friends of MAP – *Jednota přátel Masarykovy Akademie Práce* – was created with the aim of raising funds. After 1922 MAP received support from the Ministry of Education and National Culture.

15 "Dear Sirs, I have had the opportunity to reflect on the discussion that we had some time ago regarding your Academy of Work. It seems to me more and more evident that your plan to organize the entire nation's workforce represents the continuation of the wishes and notions of Comenius's Pansophia. The greatest possible output of the entire nation and the state should be founded on scientific premises; all action should constantly be based on scientific theory; a great plan should guide all the workers and unify their efforts with order and harmony. The need for a continuous project that includes every profession could not have been expressed better than in your program. And obviously it is not only a question of work, but of conscious work, work dedicated to a great goal, the moral rebirth of the nation. To explain better: even a devil can work cleverly, but the work of our people must fulfill noble objectives and be guided by an exalted ideal. It is not enough to work out of necessity, of hunger; work, or better laboriousness, is the moral duty of every conscious citizen of the Czechoslovak Republic. Love for one's neighbor, for the nation, and dedication to humanity must always be present. He who does not work, who exploits the work of others, is not Czech, is not Slovakian, is not a man. Building the republic requires great immediate and ready efforts, but beside this we have to build foundations that will reach far into future. Your Academy is making an effort to create these foundations [...]. I am attaching a check for a million crowns; this money comes from America, from our American employees and workers. I am convinced that those who have given me the responsibility of administrating these funds will approve of my decision, as I am sure that your Academy of Work will have much to learn from the American workplace." This is my translation of the original text as published in Machat, *Dějiny vědeckého řízení v kapitalistickém průmyslu*, 148, which in turn takes the document from an issue of *Nová práce* of 1920, *non vidi.* The text of the decision by the Czechoslovak Parliament can be found at: http://www.psp.cz/eknih/1918ns/ps/tisky/T2264_00.htm.

16 Masaryk, *La Nuova Europa*, 171 and Francesco Leoncini, *Nazione e minoranze nazionali nel pensiero di T.G. Masaryk,* in Francesco Leoncini, *L'Europa centrale: conflittualità e progetto* (Venice: Cafoscarina, 2003), 251-52.

17 Lawrence W. Wallace, "Industrial Management and the American Engineering Council," *Annals of the American Academy of Political and Social Science,* vol. 119, Science in Modern Industry (May 1925): 127-33.

18 Jan Padóurek, "International Contacts of the Czech Technical Academy: The Masaryk Labour Academy and the World in between the Two World Wars," *Studiae Historiae Academiae Scientiarum Bohemicae*, no. C-2 (1993): 35-50; Machat, *Dějiny vědeckého řízení v kapitalistickém průmyslu*, 143-59.

19 Václav Verunáč (1893-1960), professor at the Technical University of Prague (ČVUT), was one of the leading experts in social policy at the Academy: see Václav Verunáč, *Ústav pro technické hospodářství průmyslové* (Prague: Nákladem vlastním, 1926) and *Racionalisace, vědecká organisace a otázka sociální* (Prague: Sociální ústav republiky Českosl., 1930).

20 Jan Padóurek, "Snahy o organizování praxe československých inženýrů v USA (20 a 30 léta 20 století)," *Dějiny věd a techniky* 3 (1992): 129-39.

21 Committee on Elimination of Waste in Industry of the Federated American Engineering Societies, *Waste in Industry* (New York: Mc Graw-Hill Book Company, 1921). This study was published in Czech by MAP in 1922, under the title *Ztráty v průmyslové výrobě a jak jim čeliti*; Machat, *Dějiny vědeckého řízení v kapitalistickém průmyslu*, 150; Václav Verunáč, *Ústav pro technické hospodářství průmyslové. (První tříletí)* (Prague, 1926), 10.

22 Smrček, *Vědecká organizace práce a její aplikace ve strojírenství do konce druhé světové války*, 165-219.

23 Machat, *Dějiny vědeckého řízení v kapitalistickém průmyslu*, 148-49.

24 Frank B. Gilbreth and Lillian Gilbreth, *Motion Study* (New York: Hive, 1911); the Czech edition of this work was entitled *Studium pohybů. Metoda jak zvýšiti účinnost dělníka*, and was published in Prague in 1923 with an introduction by Stanislav Špaček. See also: Brian Price, "Frank and Lillian Gilbreth and the Motion Study Controversy, 1907-1930," *A Mental Revolution: Scientific Management since Taylor*, ed. Daniel Nelson (Columbus: Ohio State University Press, 1992), 58-76.

25 A technical-economic journal featuring social organization, techniques and politics. The journal's subtitle was: "Born on the Day of the Political Revolution (October 28, 1918) as a result of two years of effort on the part of a group of technicians in the Technical-Economic Unit that led to the creation of the most important scientific center of engineers, of the Unit of the Friends of MAP and of the Commission of State Regulation of the Capital of Prague."

26 Duccio Bigazzi, "Modelli e pratiche organizzative nell'industrializzazione italiana," in Franco Amatori et al., eds., *L'Industria*, "Storia d'Italia" Annali, vol. 15 (Torino: Einaudi, 1999), 930-33. See also Giulio Sapelli, *Organizzazione, lavoro e innovazione industriale nell'Italia tra le due guerre* (Torino: Rosenberg & Sellier, 1978), 24.

27 "Ing. Francesco Mauro první president mezinárodní delegace pro vědeckou organisaci pracovní," *Nová Práce* 8 (1926): 4; on Mauro see Paolo Viani, "Progettare l'impresa: Francesco Mauro e il dibattito europeo tra le due guerre," in *Storie d'imprenditori, ed.* Duccio Bigazzi (Bologna: Mulino, 1996), 235-92.

28 "Mario Fossati," *Zprávy Československého národního komitétu pro vědeckou organisaci*, IV (1931): 109-10.

29 Bigazzi, "Modelli e pratiche organizzative," and Duccio Bigazzi, "Organizzazione del lavoro e razionalizzazione nella crisi del fascismo 1942-1943," *Studi Storici*, no. 2 (1978): 367-96.

30 On the so-called *führerprinzip* see Mary Nolan, *Visions of Modernity: American Business and the Modernization of Germany* (Oxford: Oxford University Press, 1994); Robert R. Locke, *The Collapse of the American Management Mystique* (Oxford: Oxford University Press, 1996), 55-60; Charles S. Maier, "Between Taylorism and Technocracy: European Ideologies and the Vision of Industrial Productivity in the 1920s," *Journal of Contemporary History* 5, no. 2 (1970): 34-35; and Judith A. Merkle, *Management and Ideology: The Legacy of the International Scientific Management Movement* (Berkeley: University of California Press, 1980), 172-208. Finally, more recently, the term "technocratic internationalism" has been formulated. See: Johan Schot and Vincent Lagendijk, "Technocratic Internationalism in the Interwar Years: Building Europe on Motorways and Electricity Networks," *Journal of Modern European History*, 2 (2008): 196-216.

31 Jacques Rupnik, "The Roots of Czech Stalinism," in *Essays for Eric Hobsbawm: Culture Ideology and Politics*, eds. Raphael Samuel and Gareth Stedman Jones (London: Routledge, 1982), 304; Teichová, *The Czechoslovak Economy 1918-1980*, 65-66.

32 Lacina, *The Tempo of Economic Growth*, 207; Teichová, *The Czechoslovak Economy*, 22-23.

33 J. Rupnik, *The Roots of Czech Stalinism*, 304; Jacques Rupnik, *Histoire du parti communiste tchécoslovaque: des origines à la prise du pouvoir* (Paris: Presses de la Fondation nationale des sciences politiques, 1981).

34 *Československý národní komitét pro vědeckou organisaci* was founded in 1926 and was a spin-off of MAP, among its active members were Špaček and Verunáč.

35 Masaryk, *La Nuova Europa*, 172-73; Leoncini, *Nazione e minoranze nazionali nel pensiero di T.G. Masaryk*, XXXVI; Alfredo Laudiero, *Api e cinghiali. Culture politiche dell'Europa centrale: il caso ceco*, (Soveria Mannelli: Rubettino, 2000), 205.

36 Laudiero, *Api e cinghiali. Culture politiche dell'Europa centrale: il caso ceco*, 195-96; Francesco Leoncini, *Tomáš G. Masaryk, La Primavera di Praga e Charta 77 nell'attuale contesto politico e sociale* in *Alexander Dubček e Jan Palach protagonisti della storia europea*, ed. Francesco Leoncini (Soveria Mannelli (Cz): Rubbettino, 2009), 16-25. See also Alberto Bazala, *Il pensiero di Masaryk* (Praga: Biblioteca Internazionale di Filosofia, 1935), 33; and Eva Schmidt, *Thomas G. Masaryk's Realism: Origins of a Czech Political Concept* (München: Oldenbourg Verlag, 1984), 138-41.

37 Vladimir I. Lenin, *The Immediate Tasks of the Soviet Government*, first published on *Pravda* in April 1918; see Lenin, *Collected Works*, 4th English Edition, vol. 27 (Moscow: Progress Publishers, 1972), 235-77.

38 The novel was published in 1927. See Franz Kafka, *Amerika*, trans. Edwin Muir (New York: New Directions, 1946).

39 Archiv Akademie věd České republiky, Prague, from here on AV ČR; Masaryková akademie práce (MAP), Sig. II. 9, ser. no. 545, b. 98-101, especially files: *Návštěvy technických odborníků v ČSR* and *Stipendium Václava Klementa*. Many of the technicians who left between 1922 and 1928 were employees of L&K and Škodovy Závody (J. Černík, 1922; V. Čechura, 1926; S. Mapec, 1927; E. Kratochvíl, 1928; J. Hanuš, 1928; L. Cigánek, 1929).

40 The fellowship named after Václav Klement, founder of L&K, was assigned in 1929 to eng. Bohumír Mimra, an employee of the Tatra Kopřivnice, but it was suspended as a result of the cooling of relations after the 1929 crash.

41 Stanislav Špaček, ed., *Život a práce u Forda. Zápisky československých inženýrů z Ameriky* (Prague, 1927) and Stanislav Špaček, ed., *Ford a My. Zkušenosti československých inženýrů z americké prakse u Forda s ohledem na naše poměry* (Prague, 1928). Descriptions of the Ford industry can also be found in Vaclav Mužík, "Z prakse ve Fordových závodech v Detroitu," *Nová Práce* 3 (1925): 44-45; Jaroslav Suchánek, "Našemu autoprůmyslu," *Nová práce* 9 (1927): 74-75; K. Fast, J. Květoň and M. Lika, "Charakteristické typy fordových závodů," *Nová práce* 9 (1927): 169-75, 185-86, 204-06; Karel Hudec, "Fordův automobilismus," *Nová práce* 15 (1922): 203-07; Karel Hudec, "Henry Ford," *Nová práce* 11 and 12, 149-62. The complete reports are in AV ČR, MAP, 100.

42 "The Ford approach was to eliminate labor by machinery, not as the Taylorites customarily did, to take a given production process and improve the efficiency of the workers through a time and motion study and a differential piece rate system of payment (or some such work incentive). Taylor took production hardware as a given, and sought revisions in labor processes and the organization of work; Ford engineers mechanized work processes and found workers to feed and tend their machines. Though time and motion studies may have been employed in the setup of the machine or machine process, the machine ultimately set the pace of work at Ford, not a piece rate or an established standard for a 'fair day's work'. This was the essence of the assembly line and all the machinery that fed it. While depending upon certain elements of Taylorism in its fundamentals, the Ford assembly line departed radically from the ideas of Taylor and his followers." David A. Hounshell, *From the American System to Mass Production: 1800-1932* (London: Johns Hopkins University Press, 1985), 249-53.

43 Antonio Gramsci, *Americanismo e Fordismo* (Rome: Editori Riuniti, 1991), 42-43. See also: Karel Čapek, *RUR e L'affare Makropulos*, trans. Giuseppe Mariano (Torino: Einaudi, 1971); Jiři Fleischner, "Taylorismus," in *Technická kultura: Sociálně-filosofické a kulturně-politické úvahy o dějinách technické práce* (Prague: A. Svěcený, 1922).

44 Gramsci, *Americanismo e Fordismo*, 42.

45 Špaček, ed., *Ford a my. Zkušenosti československých inženýrů z americké prakse u Forda.*

46 Jaroslav Suchánek, "Fordův dělník," *Nová práce* 8 (1926): 168-70.

47 On the fragility of Czechoslovakian democracy, see A. Laudiero, *Api e cinghiali. Culture politiche dell'Europa centrale: il caso ceco*, 208.

48 "Použití fordových metod u nás," *Nová práce* 10 (1928): 164-65.

49 Gramsci, *Americanismo e Fordismo*, 40.

50 Bohumír Mimra, "Je možný u nás levný vůz domací výroby?" *Nová práce* 10 (1928): 162-63; Gramsci, *Americanismo e Fordismo*, 46.

51 For the consequences of the 1929 crash on Czechoslovakia, cf. Vlastislav Lacina, "The Impact of the World Economic Crisis of the 1930s on the Economy of Czechoslovakia," *Hospodářske Dějiny-*

Economic History 15 (1986): 527-43; Teichová, *The Czechoslovak Economy*, 44-46.

52 Ivan zmavc, "Jak MAP by mohla býti nápomocna k řešení úkolu národní a světové krise?" *Sborník MAP* 1 (1932): 1-3.

53 Yet, along with this, in MAP's 1930s activities it is also possible to find traces of the teachings of Elton Mayo's school of human relations. This school emphasized the need for a more complete vision of the man-company relationship, one which would consider the human factor as a variable dependent on a series of emotional factors and paying special attention to psychological needs and the importance of harmony in the workplace.

54 Václav Verunáč, "Člověk a hospodářský život," in *Encyclopedie výkonnosti* (1934): 604-17.

55 Ibid., 608.

56 Merkle, *Management and Ideology*, 172-208 and Mary Nolan, *Visions of Modernity: American Business and the Modernization of Germany*; Maier, *Between Taylorism and Technocracy*, 45-54; see also "Walther Rathenau," in *Zprávy Československého národního komitétu pro vědeckou organisaci* 6, (June 1930): 133. Walther Rathenau, *Die Neue Wirtschaft*, (Berlin: S. Fischer edition, 1918). Concerning the reorientation of the Czechoslovakian economy toward Germany which was occurring at the same time, see: Alice Teichová, *An Economic Background to Munich: International Business in Czechoslovakia 1918-1938* (Cambridge: Cambridge University Press, 1974). Besides the German influence it is also possible to see the influence of Henry Fayol and of the French debate on work organization. See, among others, Yves Cohen, *Organiser à l'aube du Taylorisme. La pratique d'Ernest Mattern chez Peugeot 1906-1919* (Besançon: Annales Littéraires de l'Université de Franche-Comté, 2001).

57 Otto Smrček, "Labor-ethics. A Czechoslovak Analogy of Technocracy," *Acta Historiae rerum naturalium necnon technicarum* 21 (Prague, 1989): 145-61; Otto Smrček, "Expanze technického myšlení," and Jan Janko, "Technokracie v Českonslovensku: několik modelů," in *Technokracie v Českých zemích (1900-1950)*, eds. Janko and Těšínská, 37-56 and 19; Alfredo Salsano, *Ingegneri e politici. Dalla razionalizzazione alla «rivoluzione manageriale»* (Torino: Einaudi, 1987), 6.

58 Leoncini, "La questione dei Sudeti come paradigma dei conflitti etnici" in *L'Europa Centrale*, 139-41 and especially the essay *Il problema delle minoranze tedesche tra le due guerre*, 117-22.

59 Leoncini, "Introduzione," in *L'Europa Centrale*; on Masaryk's "intrinsic historical correlation between democracy, socialism and nationalism and on the risk of its possible degeneration", see the observations of Andrea Graziosi, *Guerra e rivoluzione in Europa. 1905-1956* (Bologna: Il Mulino, 2001), 153-56. On Beneš' diplomacy on the eve of the Munich Pact, cf. Igor Lukeš, *Czechoslovakia between Stalin and Hitler: The Diplomacy of Edvard Beneš in the 1930s* (Oxford: Oxford University Press, 1996). After the end of the war, MAP's activity flourished again for a very short period; after 1948 it was abolished and incorporated into the Czechoslovak Academy of Sciences.

60 Tomáš Baťa, a footwear manufacturer, is considered the Czechoslovak Henry Ford. The management system that takes his name was born out of a reformulation of the principles underlying Taylor's system and his observation of American factories during a long journey to the States that he made in 1904. Not only did he finance the translation of the work "Waste in Industrial Production", but he was also a member of the presidency of MAP and participated in the activities of the Institute for Technical and Industrial Economy. The key elements of Baťa's system can be summed up as follows: daily agreements between employee and employer regarding the execution of a specific task; the relevance not of working time and the pure and simple expenditure of labor, but rather of the execution of a specific task; ban on trade union organization in the workplace; employment of non-skilled workers and fragmentation of tasks. What is more, all this was situated in a context of departmental autonomy, central to the logic behind this model. Indeed, according to Baťa, the factory would be divided into small shops which would have full organizational and managerial autonomy. Each shop would be trusted to the responsibility of a foreman and would have its own budget. Through the foreman it would periodically pay the entrepreneur a price for the use of equipment and the supply of raw materials. From week to week the foreman would come to an agreement with the entrepreneur on the price of a determined quantity of the product, whose delivery would be made at the end of the week. The risk of profit or loss would be borne by the members of the shop, and so by each single worker (*samospráva dílen*). According to Machat's analysis, this system was similar to the sweat system. The capitalist transferred the workplace, equipment and machinery and the raw materials to the foreman-entrepreneur and controlled the quality of the finished work. Others, instead, consider this model of organization both efficient and respectful of the dignity and freedom of the workers. On this question, see Hyacinthe Dubreuil, *l'Exemple de Baťa. La Libération des initiatives individuelles dans*

une Entreprise Géante (Paris 1936), who points out that, along with the introduction of machinery and new management systems aimed to provide incentives for individual production, another element that Baťa shared with Ford was the attention paid to "social activity": Machat, *Dějiny vědeckého řízení v kapitalistickém průmyslu – Vybrané kapitoly*, 183-215; AV ČR, MAP, b. 7, *Informace o ČSR pro zahraničí* and Eduard Robert Peacock, *Baťa the Henry Ford of Czechoslovakia*, 1927; "Tomáš Baťa," *Zprávy Československého národního komitétu pro vědeckou organisaci* 4 (1929): 67-68; cf. Kubů and Pátek, *Mýtus a Realita*, 269.

61 "Škodovy závody jako podnik technického a sociálního pokroku," *Nová Práce* (1925): 366.

62 Margolius, Meisl, *Škoda-Laurin and Klement* (London: Osprey, 1992), 91-93; Jaroslav Folta, ed., *Studie o technice v Českých zemích*, vol. 5 (Prague: Národní Technické Muzeum, 1918-45), 457-63.

63 František Janáček, *Největší zbrojovka monarchie. Škodovka v dějinách, dějiny ve Škodovce. 1859-1918* (Plzeň: Nakladatelství Novinář, 1990).

64 Alice Teichová and Herbert Matis, eds., *Österreich und die Tschechoslowakei: 1918-1938. Die wirtschaftliche Neuordnung in Zentraleuropa in der Zwischenkriegszeit* (Wien: Böhlau Verlag, 1996).

65 Teichová, *An Economic Background to Munich. International Business in Czechoslovakia, 1918-1938*, 203-17; Charles Beaud, "Investments and Profits of the Multinational Schneider Group, 1894-1943," in *Multinational Enterprise in Historical Perspective*, eds. Alice Teichová, Maurice Lévy-Leboyer and Helga Nussbaum (Cambridge: Cambridge University Press, 1986), 87-102. Alice Teichová has clearly shown the role of Paris in the key decisions regarding the group, above all the fact that the French influence was not based only on the possession of shares, but was also strengthened by a network of financial, contractual and personal relations. Only after the Munich Pact, and not before, did the Schneider holding sell its participation at a relatively low price to a pool of Czech banks. This was a temporary solution, as Czech capital was not strong enough to resist German pressures without foreign assistance. With 1939 and the new balance of power created by the Munich Pact in east and southeast Europe, Škoda was absorbed into the Reich's sphere of influence.

66 "Dr. Techn. H.c. Ing. Josef Havránek," in *Zprávy Československého národního komitétu pro vědeckou organisaci* 9 (November 1929): 183; Havránek began to be interested in scientific management in 1900, when he was director of the shops of the Českomoravská továrna, and afterwards he never ceased to collaborate with MAP, also taking part in the PIMCO, and collaborating with the National Committee for Scientific Management. Smrček, *Vědecká organizace práce a její aplikace ve strojírenství do konce druhé světové války*, 204-05; Václav Jíša, *Škodovy závody 1859-1865* (Prague: Práce, 1969), 80-181.

67 Karel Loevenstein, a lawyer with experience in the Wiener Bankwerein, was born in 1885 in Prague. From 1923 on he was general manager of Škoda and later became its president. He died in 1938. cf. Kubů, Pátek, *Mýtus a realita hospodářské vyspělosti Československa mezi světovými válkami*, 268-69.

68 Karlický, *Svět okřídleného šípu. Koncern Škoda Plzeň. 1918-1945*, 395-403; At Plzeň commercial vehicles, farm machinery, tanks and aircraft engines were manufactured from 1919 on. Before the L&K purchase, an attempt had been made to produce automobiles under the Balliol and Hispano Suiza license.

69 "Škodovy závody jako podnik technického a sociálního pokroku," *Nová práce* (1925): 366.

70 "Václav Klement," in *Zprávy Československého národního komitétu pro vědeckou organisaci*, no. 1 (1931): 9. Klement, who was not an engineer but a book dealer, had founded the L&K bicycle factory together with Václav Laurin. On the figure of Klement as entrepreneur, see Kubů, Pátek, *Mýtus a realita hospodářské vyspělosti Československa mezi světovými válkami*, 268-69, and for the history of the founding of the L&K, cf. Kožíšek, Králík, *L&K-Škoda*, vol. 1.

71 The ČKD, Praga was the most advanced automobile company to introduce American methods. It was also the first to introduce the assembly line. "Výroba a provoz," in *Zprávy Československého národního komitétu pro vědeckou organisaci*, no. 3 (1930): 57.

72 "Různá doporučení, daná úřady, korporacemi a jednotlivci na cestu do Ameriky," in Aša, ASAP, b. 93.

73 Letter to the *Taylor Society*, July 20, 1927 and correspondence with the *War Department*, in August of the same year in Aša, ASAP, b. 93.

74 From July 19 to August 8, New York; August 6 to 10, Connecticut (New Haven, Bridgeport, Hartford, Bullard, Pratt, and Whitney); August 10, Providence (Brown and Sharpe); August 11 to 15, Massachusetts (Worcester, The Heald Machine, Co., Springfield); August 15 to 26, New York State (Utica, Syracuse, Rochester, Buffalo, Niagara Falls); August 29 to September 6, Ohio (Cleveland); September 6 to 21, Detroit (Flint, Pontiac), then Illinois, Wisconsin, Indiana; from October 3 to 5, Cincinnati (Cincinnati Milling Machine, Le Blond, Lodge and Shippley); Washington and Philadelphia (Budd);

on specialized production in twentieth-century America see: Phil Scranton, *Endless Novelty : Specialty Production and American Industrialization, 1865-1925* (Princeton: Princeton University Press, 1997).

75 *Resumé Zprávy o studijní cestě gen.rady V. Klementa do Spojených států, vykonané spolu s Ing. J. Hauserem v době od 8-7 do 5-11-1927 k zjištění výrobních poměrů v automobilovém průmyslu americkém*, Mladá Boleslav, Febraury 5, 1928, 15 in Aša, ASAP, b. 93.

76 *Resumé Zprávy o studijní cestě gen.rady V. Klementa do Spojených států*, 7, 16.

77 *Ibid.*, 8, 9.

78 *Ibid.*, 63-66, 36-45, 52-54, 49.

79 Duccio Bigazzi, *La Grande Fabbrica. Organizzazione industriale e modello americano alla Fiat dal Lingotto a Mirafiori* (Milano: Feltrinelli, 2000), 91-93.

80 James M. Laux, *The European Automobile Industry* (New York: Twayne Publishers, 1992), 103-35; Jean-Pierre Bardou et al., *The Automobile Revolution. The Impact of an Industry* (Chapel Hill: The University of North Carolina Press, 1982), 108.

81 Karlický, *Svět okřídleného šípu. Koncern Škoda Plzeň. 1918-1945*, 396.

82 Kubů, Pátek, *Mýtus a realita hospodářské vyspělosti Československa mezi světovými válkami*, 383-84; Laux, *The European Automobile Industry*, 101.

83 Kubů, Pátek, *Mýtus a realita hospodářské vyspělosti Československa mezi světovými válkami*, 120. The authors point out that in any case the Czechoslovakian data was better than that of Belgium and Sweden; Smrček, *Dějiny*, 201.

84 Laux, *The European Automobile Industry*, 73 and Teichová, *An Economic Background to Münich*, 263-65.

85 Kožíšek, Králík, *L&K-Škoda*, vol. 2, 37.

86 Cedrych and Nachtmann, *Škoda. Auta známá i neznámá. Prototypy i seriové automobily vyráběné od roku 1934*, 48-60; Margolius, *Škoda- Laurin and Klement*, 100-03.

87 Karlický, *Svět okřídleného šípu. Koncern Škoda Plzeň. 1918-1945*, 401; Kubů and Pátek, *Mýtus a realita hospodářské vyspělosti Československa mezi světovými válkami*, 121. The first assembly line was introduced by Praga in 1928.

88 Aša, ASAP, b. 11.

89 "Zpráva," Authors: ing. Hauser, ing. Vilém Hromádko in Aša, ASAP, b. 93, May 1938.

90 "Předbežná cestovní zpráva o organisaci závodů ve Švýcarsku. Důvěrné. Organisace výroby," 10-20 July 1938, Authors: eng. E. Hirschfeld, V. Hromádko, V. Vichr in Aša, ASAP, b. 93.

91 This is only a brief mention of the German question and the role of Henlein's party; for a fuller description, see Francesco Leoncini, *La questione dei Sudeti, 1918-1938* (Padova: Liviana, 1976); and *L'Europa Centrale. Conflittualità e progetto* (Venice: Cafoscarina, 2003), 109-73; Teichová, *The Czechoslovak Economy*, 80-86; Alice Teichová, *Německá hospodářská politika v Českých zemích v letech 1939-1945* (Prague: VŠE, 1998).

92 Richard Overy, "German Multinationals and the Nazi State in Occupied Europe," in Teichová, Lévy Leboyer and Nussbaum, *Multinational Enterprise in Historical Perspective*, 301.

93 Václav Průcha, *Hospodářské a sociální dějiny Československa 1918-1992 (1. díl) Období 1918-1945* (Brno: Doplněk, 2004).

94 Teichová, *Německá hospodářská politika v Českých zemích v letech 1939-1945*, 41-47.

95 Ibid.

96 Between 1939 and 1945, 250,000 Czechoslovakian citizens were imprisoned or deported to concentration camps, Teichová, *Německá hospodářská politika v Českých zemích v letech 1939-1945*, 41.

97 Smrček, *Vědecká organizace práce a jeji aplikace ve strojírenstvi do konce druhé světoví války*, 200-05. Aša, ASAP, 12, *Cesta k firmám F. Krupp, R. Zapp, Humbold-Deutz a Elbtalwerke*, November 22-30, 1939. Authors: ing. E. Hirschfeld, Heindrych, Rádsetoulal; ing. Stachrovský; ing. Hechtberger; Pech, V. Sedláček and Aša, ASAP, 12, *Zpráva o studijní cestě po Německu*, March 25, 1941, authors: ing. E. Hirschfeld, ing. M. Scieszka, ing. Fr. Křístek; companies: Bosch, Stuttgart; Zeiss, Jena; Pittler A.G., Leipzig; F. Werner, Berlin; Hille-Werke, Dresden; Elbewerke, Dresden.

98 Smrček, *Vědecká organizace práce a její aplikace ve strojírenství do konce druhé světové války*, 219; Correspondence between the Českomoravský Národní Komitét pro Vědeckou Organisaci and the Masarykova akademie práce between November 1939 and May 1940 in AV ČR, MAP, b. 99.

99 Smrček, *Vědecká organizace práce a její aplikace ve strojírenství do konce druhé světové války*, 205.

100 Teichová, *An Economic Background to Munich*, 217.

101 Cf. Overy, *German Multinationals and the Nazi State in Occupied Europe*, 301.

102 Kožíšek and Králík, *L&K-Škoda*, vol. 2, 67-78.
103 Aša, ASAP, 4, Mladá Boleslav, June 12, 1965.
104 "Oběžník generálního ředitele," no. 345, October 13, 1942 in Aša, ASAP, b. 3.
105 Bigazzi, *La grande fabbrica*, 103-09, see also Bigazzi, *La fabbrica nella crisi del regime fascista*, 215-79.
106 *Seznam zaměstnanců, kteří byli zatčení Gestapem, Dělnický referát*; *Seznam zaměstnanců, kteří zemřeli při odzbrojování německé armády ve dnech 5.5. až 7.5.1945*, *Seznam zaměstnanců, kteří zemřeli při náletu dne 9.5.1945 nebo následkem zranění* in Aša, ASAP, b. 3.

Chapter 2

1 Victor S. Mamatey and Radomír Luža, eds., *A History of the Czechoslovak Republic, 1918-1948* (Princeton: Princeton University Press, 1973); in particular the study by Radomír Luža, *Czechoslovakia between Democracy and Communism, 1945-1948*, 387-415; Karel Kaplan, *Pravda o Českonslovensku, 1945-1948* (Prague: Panorama, 1992); and by the same author: *The Short March: The Communist Take-over in Czechoslovakia 1945-1948* (London: C. Hurst, 1987); and *Pět Kapitol o únoru* (Prague: Ústav pro soudobé dějiny, 1996). On Czechoslovakia's diplomatic relations with European nations and with the USSR, see the more recent *Československo v poválečné Evropě* (Prague: Karolinum, 2004).

2 Bradley Abrams, *The Struggle for the Soul of the Nation: Czech Culture and the Rise of Communism* (Oxford: Rowman and Littlefield Publishers, 2004), 88-138, 156-77; this author analyzes the writings of the main Czech intellectuals in the years immediately after the war, giving a very convincing representation of the political climate and culture of the period 1945-48. See also Jan Gross, "The Social Consequences of War: Preliminaries for the Study of the Imposition of Communist Regimes in East Central Europe", *East European Politics and Societies* 3 (1989): 198-214; Istvan Bibò, *Misère des petits états d'Europe de l'Est* (Paris: L'Harmattan, 1986).

3 The Czechoslovak Communist Party was born in 1921; its Bolshevization coincided with the assumption of leadership by Klement Gottwald in 1929, which led to a radical change in the way of relating to work and to the worker, who was transformed from "ouvrier citoyen" to "prolétaire bolshevique", according to the definition of Jacques Rupnik, *Histoire du parti comuniste*, 84. Rupnik reports that this change marked a profound estrangement of the working-class movement from the Communist Party and a significant loss of Party members: of the 210,000 members in 1924, only 113,000 remained in 1930. It is also worthy to note that in the 1920s, despite the fact that it was the fourth largest in the world for number of enrolled workers after Germany, Great Britain and the U.S., the Czechoslovak trade union movement was highly fragmented according to craft, party, and even nationality, and its action was more reformist than revolutionary. The Communist Party, though politically strong, was quite weak among unionized workers, with only 11 per cent of unionized workers being members. During World War II, the trade union movement found united representation for the first time in the URO (*Ústřední rada odborová*), founded in 1943 as the coordinating organ for the revolutionary union struggle. The URO played a key role in the Prague insurrection and in general in what was defined as *odopor*, the passive resistance by workers to the Germans (in contrast to *odboj*, the active resistance of emigrated intellectuals). Unity of the workers' movement became a key word in trade unionism in the years 1945-48, along with the motto *jeden závod, jedna odborová organisace* [one factory, one organization]. The passage was therefore made to industrial unionism, in which factory representatives were democratically elected without any political or craft distinction and answered directly to the URO. In 1945 the URO founded the *Revoluční odborové hnutí* [ROH; Revolutionary Union Movement], whose base was made up of the workers' councils created in every factory. Again according to Rupnik, based on the factory councils, a "vague de radicalisme dans la classe ouvrière" became widespread; workers took on a very important role in company management and strongly supported nationalization. Rupnik further notes that the relationship was immediately tense between the Czechoslovak Communist Party, which looked on the union essentially as a conveyor belt between the Party and the masses, and the workers' councils. In the three years from 1945 to 1948, a series of political measures aiming at institutionalizing the factory councils was taken. However, the price for their institutionalization was an attempt by the Party to subordinate them, thereby undermining their independence. The next step, in 1948, was their elimination, as they served no purpose in the workers' state. See also Jan Bloomfield, *Passive Revolution: Politics and the Czechoslovak Working Class 1945-1948* (London: Allison and Busby), 68-105. On the profound de-politization of the working class

that resulted, Jiří Valenta also writes in "Czechoslovakia: a Prolétariat Enbourgeoisé?" in *Blue Collar Workers in Eastern Europe*, eds. Jan F. Triska and Charles Gati (London: George Allen and Unwin, 1981), 209-23.

4 The post-war economic situation is well summed up in the study by Jan M. Michal, "Postwar Economic Development," in Mamatey and Luža, *A History of the Czechoslovak Republic, 1918-1948*, 428-60. Also very clear on the transformation in Czechoslovak industry and the process of concentration is: Alice Teichová, "The Halting Pace to Scope and Scale" in *Big Business and the Wealth of Nations*, eds. Alfred D. Chandler, Franco Amatori, Takashi Hikino (Cambridge: Cambridge University Press, 1997), 447-61. See also Alice Teichová, *The Czechoslovak Economy 1918-1980*; and Derek H. Aldcroft and Steven Morewood, *Economic Change in Eastern Europe since 1918* (Aldershot: Edward Elgar, 1995), 90-91.

5 Teichová, *Czechoslovak Economy*, 89-93.

6 Michal, *Postwar Economic Development*, 438-45.

7 Some data in François Fejto, *Storia delle democrazie popolari* (Milan: Bompiani, 1978), 125: "70 billion crowns, of which 36 percent destined for industry and mines, 22 percent to transport, 35 percent to building homes, 7 percent to agriculture. The plan was brought to a successful conclusion: production of coal, electricity, machines, iron and steel was 10 percent higher than in 1936. Only in the textile industry and in agriculture were the estimates not reached." (or. François Fejto, *Histoire des démocraties populaires. Tome 1: l'ère de Staline*, Paris: Editions du Seuil, 1969).

8 Fejto, *Storia delle democrazie popolari*, 100. The author quantifies UNRRA aid as 261 million dollars, managed by an organization placed under the direction of František Němec. These funds were used to reconstruct agro-technical assets and some of the industrial equipment. For a fuller treatment of the reasons underlying Czechoslovakia's refusal of the Marshall Plan, see: Alice Teichová, "Les hésitations tchèques sur le Plan Marshall," in *Le Plan Marshall et le relèvement économique de l'Europe. Colloque tenu à Bercy les 21, 22, et 23 mars 1991*, eds. René Girault and Maurice Lévy-Leboyer (Paris: Comité pour l'histoire économique, 1993), 109-18. See also in the same collection Mikhail Narinski, *Le Plan Marshall et l'URSS*, 119-23; and Zdeněk Veselý, *Československo a Marshallův Plán*, (Prague: Acta Universitatae Carolinae Philosophica et Historica Monographia XCII, 1982).

9 Teichová, "Les hésitations tchèques sur le Plan Marshall," 112-14.

10 Luža, *Czechoslovakia between Democracy and Communism*, 409-15.

11 Kožíšek, *L&K-Škoda*, 91. In the late 1920s the owner of the Jawa motorcycle factory bought some land near Solnice, some of which was occupied by a sawmill and other industrial plants; he turned them into an important plant for building car bodies, where starting from 1934, the bodies for the Jawa 700 and Jawa Minor were produced. Every day four bodies were completed and then sent by rail to Týnec nad Sázavou, where assembly was carried out. During the Second World War, the prototypes of the Jawa Minor II were built. After its merger with AZNP in 1947, this plant produced limited series of the Škoda Superb, ambulances, and the Škoda 1200 and 1201.

12 Státní Ůstřední archiv (from here on SÚA), Československé závody kovodělné a strojírenské (henceforth Čzks), *Úvod*. See also Kožíšek, *L&K-Škoda*, vol. 2, 67-78; and Martin Myant, *The Czechoslovak Economy 1948-1988. The Battle for Economic Reform* (Cambridge: Cambridge University Press, 1989), 18-30.

13 See the studies of Alfred Dupont Chandler, Jr., *Strategy and Structure: Chapters in the History of Industrial Enterprise* (Cambridge: Massachusetts Institute of Technology Press, 1962); and by the same author, *The Visible Hand: The Managerial Revolution in American Business* (Cambridge: Belknap Press, 1977). See also: Pier Angelo Toninelli, ed., *The Rise and Fall of State-Owned Enterprise in the Western World* (Cambridge: Cambridge University Press, 2000).

14 Aša, Automobilové závody, národní podnik, 1945-1948 (henceforth AZNP/P), b. 4 "Akce Taub," Alexander Taub, *A People's Technology. A Report to Dr. Ing. F. Fabinger, General Director of Czechoslovak Metalworking and Mechanical Engineering Plants*, Prague, September 1946; see also the other two reports: Alexander Taub, *Automobilový Průmysl pro Československo*, 1947 and Taub, *3. Zpráva, I. Program výrobků, II. Závody a zařízení*, 1947.

15 Aša, AZNP/P, b. 4 "Akce Taub," Jaroslav Frei, *Zpráva o prohlídce amerických automobilových a motocyklových závodů v době od 10. do 22. června 1946.*

16 See especially: Charles Maier, "The Politics of Productivity: Foundation of American International Policy after World War II," *International Organization* 31 (September 1977): 607-33 and Charles Maier, *In Search of Stability: Explorations in Historical Political Economy* (Cambridge: Cambridge University Press, 1984); in Italian in Charles Maier, *Alla ricerca di stabilità* (Bologna: Il Mulino, 2003),

177-222 (*In Search of Stability: Explorations in Historical Political Economy*, Cambridge: Cambridge University Press, 1987); see also the contributions of Jacqueline McGlade, "Lo zio Sam ingegnere industriale. Il programma americano per la produttività e la ripresa economica dell'Europa occidentale (1948-1958)," *Studi Storici* 37, no. 1 (January-March 1996): 9-40; "Americanization: Ideology or Process? The Case of the United States Technical Assistance and Productivity Programme," in Zeitlin, *Americanization and its Limits,* 53-75; "The US Technical Assistance Program: From Revolutionary Vision to Production Drive," in *Catching up with America: Productivity Missions and the Diffusion of American Economic and Technological Influence after the Second World War*, ed. Dominique Barjot (Paris: Presses de l'Université de Paris-Sorbonne, 2002), 67-86. Scholars have extensively debated on the extent, goals, methods, and impact of the American intervention in the reconstruction of Western Europe. See for example Alan Milward, *The Reconstruction of Western Europe, 1945-1951* (London: Routledge, 1984); Paul Hogan, *The Marshall Plan: America, Britain, and the Reconstruction of Europe* (Cambridge: Cambridge University Press, 1987).

17 The OEM was founded in the Executive Office of the President by administrative order on May 25, 1940. It housed a series of civil agencies whose job it was to manage the war effort. See Maier, *Alla ricerca di stabilità*, 178; and Olivier Zunz, *Perché il secolo americano?* (Bologna: Il Mulino, 2000), 137 (or. *Why the American Century?*, Chicago-London: The University of Chicago Press, 2000); Also see John Lewis Gaddis, *The United States and the Origins of the Cold War, 1941-1947* (New York: Columbia University Press, 1972).

18 I owe this information to a personal communication from Jennifer Taub; it is confirmed in "Foreign Trade: China's Plan," *Time*, January 29, 1945. "China's Plan was born late in 1943, when Lauchlin Curie, then acting deputy administrator of the Foreign Economic Administration, sat down to discuss it with FEA's electric, little, British born chief engineer, Alex Taub. Taub had just completed a ten-year industrialization plan for Brazil."

19 Missouri Historical Society, from the William McChesney Martin Jr. Collection, Memorandum for Mr. Sidney Sherwood, January 21, 1946.

20 Maier, *Alla ricerca di stabilità*, 177.

21 Taub, *A People's Technology,* 1.

22 Ibid., 6.

23 Ibid., 3.

24 Ibid., 3.

25 Ibid., 82. Škoda's retail price was 70,000 crowns, equivalent to 75 crowns per kilogram, since it weighed 936 kgs; the Jawa also cost 75 crowns per kilogram, as its price was 52,000 crowns and its weight 685 kg; instead, the Tatra "87" which cost 160,000 crowns and weighed 1,500 kg had a per kilogram cost of 106 crowns.

26 Ibid., 80.

27 Taub's reasoning was as follows: in Great Britain pre-war production had reached four hundred thousand vehicles per year; it would get back to this volume in the course of 1947, and it was foreseeable that British industry would soon reach eight hundred thousand units per year. It would take France a little longer to reach its pre-war volume of two hundred thousand units per year, but Taub also foresaw that French production would reach six hundred thousand units. By contrast, German industry would not regain its pre-war level of three hundred thousand autos per year: it was here that as a newcomer, Czechoslovak industry could assert itself.

28 Taub, *A People's Technology,* 2.

29 Ibid., 68.

30 Ibid., 43.

31 Ibid., 40.

32 Ibid., 12.

33 Ibid., 13.

34 Ibid., 60-70.

35 Ibid., 44.

36 Ibid., 62.

37 Ibid., 34.

38 Abrams, *The Struggle for the Soul of the Nation*, 109. The Communist intellectuals insisted above all on the physical dimensions of state power. On the "malost": Tomáš Garrigue Masaryk, *Problém malého národa* (Prague: Neutralita, 1990); and by the same author, *Česká otázka* (Prague: Svoboda, 1990).

39 "To meet this economic situation, the Czechs must operate their automotive industry as if it were one manufacturing unit. They must centralize, specialize and standardize – a central plant for engines, transmission and axle-differential, body and sheet metal and accessories. The principle of separation and concentration is the most important factor for the success of the Czechoslovak industry. The automotive industry must be concentrated and all products not part of that industry must be removed or separated from it. It is not possible to operate effectively at the tempo required by this industry if operations are mixed with parts that are out of character", in Taub, *A People's Technology*, 34.
40 "General Motors supports some relatively low producers for America. However, they have the advantage of many items that are common to all and these common parts are made in central plants. Without this, the low producers in GM would be in serious trouble. [...] This principle of separation and concentration prevails throughout the industry and in such concerns as General Motors, we find many factories operating as central plants. The main plants of Buick, Chevrolet, Pontiac, Oldsmobile and Cadillac all draw about half of their parts from central plants. This is necessary for minimizing development and manufacturing costs" (Ibid., 15 and 35).
41 Ibid., 70.
42 Ibid., 3.
43 Ibid., 36.
44 Ibid., 71.
45 Ibid., 80.
46 Ibid., 84.
47 Taub, *Automobilový průmysl pro Československo*: this report reiterates many of the same concepts as the previous one; there is, however, greater emphasis on the fundamental importance of the automobile industry for the nation, and without doubt there is more rhetoric, 2-4 and 17.
48 Ibid., 40.
49 Ibid., 41.
50 Taub, *Zpráva, I. Program výrobků, II. Závody a zařízení*, 1947.
51 Ibid., 2.
52 Hounshell, *From the American System of Manufacturing to Mass Production*, 264; and Tolliday, ed., *The Rise and Fall of Mass Production*; see also the following articles in the special issue of *Enterprise and Society* 1 (2000): Jonathan Zeitlin, "Reconciling Automation and Flexibility? Technology and Production in the International Automobile Industry"; Mark Schwartz, "Markets, Networks, and the Rise of Chrysler in Old Detroit, 1920-1940"; David A. Hounshell, "Automation, Transfer Machinery, and Mass Production in the US Automobile Industry in the Post-World War II Era".
53 Robert Locke, *The Collapse of the American Management Mystique* (Oxford: Oxford University Press, 1996), 36-54.
54 Between the last years of the 1990s and the first years of the new millennium, European and American historiography thoroughly explored the concrete and ideological aspects of the transfer and reception of the American model in western European managerial and technical environments. Besides the aforementioned Bigazzi, Zeitlin and Herrigel, see also Matthias Kipping and Ove Bjarnar, eds., *The Americanization of European Business: The Marshall Plan and the Transfer of US Management Models* (London-New York: Routledge, 1999).
55 Locke, *The Collapse of the American Management Mystique*, 10.
56 Maier, *In cerca di stabilità*, 179.
57 Zeitlin, "Introduction," in Zeitlin, *Americanization and its Limits*, 5.
58 Aša, AZNP/P, b. 4, "Soubor přednášek o poznatcích získaných v USA přednášeno v AKRCs dne 2.února 1948": František Fabinger, *Zahájení přednášek automobilových odborníků po návratu z USA*, 1948; Vladimír Matouš, *Poznatky z americké konstrukce osobních vozů*; Václav Křěmár, *Výrobní pomůcky používané v Usa;* Richard Kneschik, *Příprava výroby a tovární zařízení v automobilkách USA s ohledem na použití získaných poznatků pro náš průmysl*; Zdeněk Kejval, *Karoserie v USA*. See also the complete reports: Vladimír Matouš, *Cestovní zprávy z USA, Výtah z cestovních zpráv od 31. srpna do 12. listopadu 1947*; Václav Křěmár, *Automobilové továrny v USA, Zpráva z cesty konané v červenci-září 1947*. Others reports bear the signatures of Marcel Fremunt, Augustin Zemene, Karel Výška.
59 Vladimír Matouš was born in 1896 in Liberec and died in 1963; he graduated from the *České vysoké učení technické* (ČVUT) [Technical University of Prague] in 1920. He worked first for Walter in Jinonice and later at the Škoda shops in Plzeň, when they were attempting to produce the Hispano Suiza on license. In 1928 he was transferred to Mladá Boleslav, where he remained until 1959, at first as

director of automobile design production. From 1936 he served as vice-director of ASAP, and visited the US on several occasions. In 1947 Minister Bohumír Laušman named him expert in the Ministry of Industry; in 1948 he was once again at AZNP as technical director, and in 1950 he became director of technical development. Under his direction the Škoda 1101, 1200 and 440 (Spartak) were designed; see: Kožíšek, *L&K-Škoda*, 22, 32, 122; and Jan Králík, *V soukolí okřídleného šípu. Zákulisí automobilky Škoda očima konstruktéra Favoritu Petra Hrdličky* (Prague: Grada Publishing, 2008), 19.

60 Zdeněk Kejval was born in 1905 and died in 1988. He graduated from the Technical School of Mechanical Engineering in Plzeň. He began as a body designer for the Praga, and later, in 1935, became director of the Kvasiny factory, where he designed the body of the Jawa. In 1947 he was sent to America and also to Germany. From 1947 to 1954 he was the technical director for the Praga, where he supervised press manufacture. In 1955 he went to AZNP, where he was responsible for planning; from 1961 he was director of the body and small press shops and of the foundry; in this capacity he also coordinated part of the construction of AZNP's new plant for the MB1000. He also gave courses on planning and design at the Faculty of Mechanical Engineering, for the Automobile and Tractors Chair of the ČVUT: Kožíšek, *L&K-Škoda*, 40, 84, 153; and Králík, *V soukolí okřídleného šípu*, 48; Václav Křěmár was responsible for the preparation of production at Mladá Boleslav.

61 František Žalud was born in 1919 and died in 1993; he graduated from the University of Strathclyde in Glasgow, Scotland. Afterwards, he worked as an assistant at British Polar Engines. He returned to Czechoslovakia in 1946 and worked at ČKD in the machine tool research department, before going to *Ústav pro výzkum motorových vozidel* (ÚMVT) [the research institute for motor vehicles], founded in 1952, where he remained until 1988; in Králík, *V soukolí okřídleného šípu*, 38; and his recollections: František Žalud, *Přežili jsme* (Prague: Trilobit, 1998), 60-64.

62 Aša, AZNP/P, b. 4, V. Křěmár, *Automobilové továrny v Usa*; Kneschik, *Příprava výroby*.

63 Frei, *Zpráva*, 4.

64 Křěmár, *Automobilové továrny v Usa*.

65 Frei, *Zpráva*, 4.

66 Aša, AZNP/P, 4, Matouš, *Poznatky z americké konstrukce osobních vozů*. Cf. Zunz, *Perchè il secolo americano*, 172-75.

67 Ibid., 2.

68 Ibid., 6.

69 Ibid., 5.

70 J. Frei, *Zpráva*, 6

71 Aša, AZNP/P, b. 4, *Soubor přednášek o poznatcích získaných v USA přednášeno v AKRCs dne 2.února 1948.*

72 Aša, AZNP/P, b. 4, *Akce Renault*: Václav Beneš, *Cestovní zpráva ze studuijní cesty u firmy Renault, Francie*, April 1947; there were also the reports of Ladislav Krahulík, director of body manufacturing and of Josef Velebný, body designer, who wrote the report made for the Mostra della carrozzeria italiana held in Milan in November, 1947.

73 See especially the final report by Jaroslav Frei: "Cesta z těžke situace našeho průmyslu motorových vozidel jest: najíti správný způsob aplikace všech poznatků z Ameriky československým způsobem, abychom co nejdříve dosáhli evropské úrovně" in Aša AZNP/P, b. 4, *Soubor přednášek o poznatcích získaných v USA přednášeno v AKRCs dne 2.února 1948.*

74 Beneš, *Cestovní zpráva ze studijní cesty u firmy Renault, Francie*, 18.

75 See Duccio Bigazzi, "L'ora dei tecnici: aspirazioni e progetti tra guerra e ricostruzione," in Giuseppe De Luca, *Pensare l'Italia nuova: la cultura economica milanese tra corporativismo e ricostruzione* (Milan: Franco Angeli, 1997), 379-433.

76 Frei in Aša, AZNP/P, b. 4, *Soubor přednášek o poznatcích získaných v USA přednášeno v AKRCs dne 2.února 1948.*

Chapter 3

1 Karel Kaplan, *Českonslovensko a RHVP (1948-1953)* (Praha: Ústav pro soudobé dějiny AV ČR, 1995).
2 Myant, *The Czechoslovak Economy 1948-1988*, 31.
3 Karel Kaplan, *Komunistický režim a politické procesy v Československu* (Brno: Barrister-Principal, 2001).
4 *Státní úřad plánovací*: the institute was created in August 1945 by President Beneš (decree 63/1945). Its functions were modified by law in March 1949 and again in November 1951. In 1959 the office was eliminated and its functions taken on by the *Státní plánovací komise* [State Planning Committee].
5 Teichová, *The Halting Pace to Scale and Scope*, 447-56.
6 *Úvod*, in SÚA, ser. Československé závody automobilové a letecké (henceforth Čzal).
7 *Organisace závodního ředitelství*: spec. organizational chart, 29 April 1949, in Aša, AZNP/P, b. 12 and *Oběžník podnikového ředitele č.3*, 21 May 1949, in Aša, AZNP/P, b. 14.
8 *Zápis z pracovní s prozatímně pověřenými vedoucími podniku ČZAL v Praze*, 20 September 1949, in Aša, AZNP/P, b. 1.
9 Wlodizimier Brus, "1950 to 1953: The Peak of Stalinism," in *The Economic History of Eastern Europe 1919-1975, vol. III: Institutional Change within a Planned Economy*, eds. Michael Charles Kaser and Edward A. Radice (Oxford: Clarendon Press, 1985), 3-39; Teichová, *The Czechoslovak Economy*, 134-40.
10 Brus, "1950 to 1953: The Peak of Stalinism," 14-15.
11 Mathias Roeser, "Sovětizace a militarizace československého leteckého průmyslu v letech 1949-1953," *Soudobé dějiny*, 1, (1996): 60-71.
12 Files: *Organisační předpis podnikového ředitele* and especially that of October 4, 1952, no. 2-52 and those relative to *Reorganisace řízení*, no. 3-52, 4-52 in Aša, ser. Automobilové závody, národní podnik, 1948 (henceforth AZNP), b. 11.
13 "Nemístní luxus," *Svět motorů* 76 (1950): 164.
14 *Zapis z pracovní porady s prozatímně pověřenými vedoucími podniku ČZAL v Praze*, September 20, 1949, in Aša, AZNP/P, b. 1. Similar reasoning can also be found in minutes of the meeting with the general director Horák at SÚP, August 30, 1949, in SÚA, Čzal, b. 83.
15 Minutes of the business meeting with the temporary managers of the Čzal enterprises, September 20, 1949: "The population," wrote the general director of Čzal, eng. Horák to the SUP, "still wants the people's car, there is domestic and foreign demand."
16 *Problematika výroby automobilů a motocyklů*, 1949 in SÚA, Čzal, b. 86.
17 Ibid.
18 "Zvýšíme brannost našeho národa," *Svět motorů* 116 (1951): 1443; "Poslání a úkoly československych motoristů," *Svět Motorů* 6 (1952).
19 Adolf Tůma, "Rozvoj motorové dopravy," *Svět motorů* 83 (1950): 387, the author was president of the Autoclub.
20 Ibid. See also Adolf Tůma, "Reorganisace autoklubu RČS a jeho příští úkoly," *Svět Motorů* 74 (1950): 99. The author continued, "we know from the complaints made to our magazine how many people would like to drive a car, how many want a people's car to be produced, how many ask for car prices to be reduced, like in the West. True, it's done. But the question is for whom it's done and in whose favor!"
21 Tůma, "Reorganisace Autoklubu RČS a jeho příští úkoly," 99.
22 Tůma, "Rozvoj motorové dopravy," 387.
23 "Reorganisace Mototechny," *Svět motorů* (1952): 97; Mototechna was the agency in charge of distribution on the domestic market, Motokov was the distributor for the foreign market.
24 *Program vývojového oddělení pro vozidla* document signed by engineer Karel Výška, June 22, 1949, in SÚA, Čzal, b. 64.
25 *Český statistický úřad, zprávy a rozbory, rozvoj automobilového průmyslu v ČSR* in Aša, AZNP, b. 34, 1970.
26 *Seznam obráběcích strojů k disposici na stroje výkonnější, odpovídající požadovanému zvýšenému počtu vozů dle programů* in Aša, AZNP, b. 26.
27 *Seznam obráběcích strojů které potřebují AZNP, v MB pro 5 letý plán v roce 1949-1953* in Aša, AZNP, b. 26. For a comparison with other European manufacturers, see Franco Amatori, *Impresa e Mercato. Lancia, 1906-1969* (Bologna: Il Mulino, 1996), 154; Bigazzi, *La grande fabbrica*, 130-45; Nicolas Hatzfeld, *Les gens d'usine. 50 ans d'histoire à Peugeot-Sochaux* (Paris: Les Editions de l'Atelier, 2002), 246-47.

28 Minutes of the management, October 28, 1948, Aša, AZNP, b. 1.
29 Ibid.: Eng. Matouš complained that there was no remedy for the loss attributed to the *AZNP*: the prices of materials had risen, while the AZNP could not raise the price of its vehicles either on the domestic or the foreign market. Increasing the quota for export would not help the situation. The disproportion between what the state earned with the export of cars and what it reinvested in the sector was well known.
30 Jan Králík, *V soukolí okřídleného šípu. Zákulisí automobilky Škoda očima konstruktéra Favoritu Petra Hrdličky* (Prague: Grada Publishing, 2008), 47-48. Petr Hrdlička was the son of Karel Hrdlička, director of the Mladá Boleslav factory ASAP until 1941, and forced to retire after the war, since he was not a Party member.
31 *Dotazy pro studijní cestu v SSSR e rozbor otázek pro SSSR. Automobilové závody nár. podnik, závod gen. L. Svobody, Mladá Boleslav*, September 11, 1950, in Aša, AZNP/P, b. URSS.
32 Teichová, *The Halting Pace to Scale and Scope*, 435 (in the years between 1950 and 1971, 78.5 percent of the growth in the blue-collar workforce was represented by women).
33 Zdeněk Kejval, *Karoserie v USA*, in Aša, AZNP/P, b. 4: the most important progress was in body manufacture, in intra-shop transport and, above all, in the technology used for finishing and precision work (synthetic paints had substituted nitrocellulose, lapping, and the technique of superfinishing had been introduced to replace honing; the error leeway reached by new-generation Natco-type grinders and borers was minimal).
34 *Rozbor otázek pro SSSR.*
35 Ibid.
36 Ibid.
37 Ibid., questions from 157 to 164.
38 Ibid., questions from 165 to 166.
39 Ibid., questions from 165 to 166.
40 Ibid., questions from 165 to 166.
41 *Konference závodních ředitelů, Zápis ze dne 4-1-1951, 1-2-1951, 1-3-1951, 5-4-1951, 3-5-1951, 31-5-1951,* monthly reports until November 29, 1951 and for all of 1953 in Aša, AZNP, b. 7.
42 *Program vývojového oddělení pro vozidla*, June 22, 1949, in SÚA, Čzal, b. 64.
43 *Vědecko-technická spolupráce s SSSR*, minutes of the meetings of the Committee for the technical scientific cooperation from December 1948 to March 1949, in SÚA, Čzal, b. 83.
44 *Spolupráce SSSR, Československé závody na výrobu vozidel. Zprávy ze studijní cesty ČS. Automobilových závodů*, February 9-April 9, 1951, in SÚA, Čzal, 27. Cf. Lewis Siegelbaum, *Cars for Comrades: The Life of the Soviet Automobile* (Ithaca: Cornell University Press, 2008).
45 *Vědecko-technická spolupráce s SSSR*; "Motorisace Sovětského Svazu," *Svět motorů* 91(1950): 643; "Motorismus v Sovětském svazu," *Svět motorů* 95 (1951): 786; "Sovětský automobilní průmysl," *Svět motorů* 98 (1951): 870.
46 *Program vývojového oddělení pro vozidla.*
47 Ibid.
48 Ibid.; See Chandler, *Dimensione e diversificazione. Le dinamiche del capitalismo industriale,* 565; Richard Whittington and Michael Mayer, *The European Corporation: Strategy, Structure, and Social Science* (Oxford: Oxford University Press, 2000).
49 *Program vývojového oddělení pro vozidla.*
50 "If we were to compare the entire Czechoslovak automotive industry to a large American corporation, we would see that we have all that is needed for common management of its development. What the English interest groups have managed to do for auto products has been made possible by the fact that they share their offices, they use the same or similar engines, gearboxes and rear axles, common to all the allied companies, they see standardization through to the end, they concentrate production and share a common sales policy: this is the reason for the success they have today. The Americans use parts and equipment made out of standardized parts to an incredible extent, without caring about the fact that they are used both on the most economic and the most expensive cars." Ibid.
51 Ibid.
52 Ibid.
53 Janos Kornai, *The Socialist System: The Political Economy of Communism* (Oxford: Oxford University Press, 1992), 97-109. See also Neil Fligstein, *The Transformation of Corporate Control* (Cambridge: Harvard University Press, 1990) and Robert F. Freeland, "The Myth of the M-Form? Governance,

Consent, and Organizational Change," *American Journal of Sociology* 2 (1996): 483-526; Robert F. Freeland, *The Struggle for Control of the Modern Corporation: Organizational Change at General Motors, 1924-1970* (Cambridge: Cambridge University Press, 2001).

54 *Program vývojového oddělení pro vozidla.*

55 Ibid. and *Některá hospodářská porovnání mezi USA e ČSR, eng. František Žalud* in SÚA, Čzal, 64; see also Alfred Chandler, *Strategy and Structure: Chapters in the History of the Industrial Enterprise* (Cambridge: Massachusetts Institute of Technology, 1966).

56 Werner Abelshauser, "Two Kinds of Fordism: On the Differing Roles of the Automobile Industry and the Development of the Two German States," in *Fordism Transformed, the Development of Production Methods in the Automobile Industry*, ed. Haruito Shiomi and Kazuo Wada (Oxford: Oxford University Press, 1995).

57 Kaplan, *Československo a RHVP (1948-1953)*; Michael Charles Kaser, *Comecon: Integration Problems of the Planned Economies* (London: Oxford University Press, 1965).

58 Václav Průcha, *Sectoral Changes in Industry after World War II* (Prague: Institute of History of the ČSAS, 1991); Teichová, *The Czechoslovak Economy 1918-1980*, 134-49.

59 *Informace o zahraničním obchodu ČSR s kap. státy za r. 1952 ve srovnání s r. 1950 a 1951*, in SÚA, ser. Hospodářská rada, ůstřední výbor KSČ (from here on HR ÚV KSČ),43, 305. See also Bohumír Brom and Eduard Kubů, "The Role of the Czechoslovak Trade with Neutral Countries in the Period of Escalating Cold War: The Example of Switzerland and Sweden 1949-1953" and Dagmara Jajesniak-Quast, "Permeating the Embargo with Iron and Steel: Poland, Czechoslovakia, the GDR and the Neutral States," in *Gaps in the Iron Curtain: Economic Relations between Neutral and Socialist Countries in Cold War Europe*, eds. Gertrude Burcel-Enderle, Alice Teichova and Piotr Franaszek (Warsaw: Jagiellonian University Press, 2008), 165-91 and 270-89.

60 In 1949 the USA created the Coordinating Committee for Multilateral Export Control (COCOM), which was integrated by secret bilateral agreements aiming at preventing East-West trade. On the effectiveness of these agreements, see Gunnar Adler-Karlsson, *Western Economic Warfare 1947-1967. A Case Study in Foreign Economic Policy* (Uppsala: Almqvist, 1968) and the contributions included in *East West Trade and the Cold War*, eds. Jari Oyala and Jari Eloranta, (Jyväskylä: Jyväskylä University Printing House, 2005).

61 *Informace o zahraničním obchodu ČSR s kap. Staty za r. 1951 ve srovnání s r. 1950 a 1951* in SÚA, HR ÚV KSČ, 43, 305.

62 *From Fabinger to Frejka*, February 21, 1949, in SÚA, HR ÚV KSČ, 43, 306.

63 *From Frejka to Fabinger*, March 27, 1949, in SÚA, HR ÚV KSČ, 43, 306. See also especially the Czechoslovak delegate's reports to the Bureau of the Comecon of February 23 and 28, 1949, also in SÚA, HR ÚV KSČ, 43, 306.

64 *Zahraniční obchod Československa v prvním poletí 1949 and Přípravný materiál pro presidium HR v Požarech*, April 27, 1949, in SÚA, HR ÚV KSČ, 44, 306.

65 *From Goldmann to Frejka* between February and April 1949 in SÚA, HR ÚV KSČ, 43, 306.

66 *O vědecko-technické spolupráci*, March 27, 1949, signed by Fabinger in SÚA, HR ÚV KSČ, b. 43, 306.

67 *O vědecko-technické spolupráci*, Fabinger was in fact very inflexible as regards the reciprocity of obligations among the people's democracies.

68 *Státní úřad plánovací, Zápis z porady dne 30-8-1949 na SÚP* and *Navštěva u s. Ing Maryšky na SÚP, dne 13-4-1950 ve 2 hod. odpoledne* in SÚA, Čzal, b. 83; *Výroba automobilů v ČSR v ramci spolupráce se SSSR a zeměmi lidových demokracií*, April 12, 1950, in SÚA, Čzal, b. 83.

69 *Problematika výroby automobilů a motocyklů.*

70 Ibid.

71 Ibid.

72 Ibid.

73 *Státní úřad plánovací, Zápis z porady dne 30-8-1949 na SÚP.*

74 Ibid.; Hounshell, *From the American System to Mass Production*; David Hounshell, "Planning and Executing 'Automation' at Ford Motor Company, 1945-1965: The Cleveland Engine Plant and Its Consequences," in *Fordism Transformed: The Development of Production Methods in the Automobile Industry*, eds. Haruhito Shiomi and Kazuo Wada (Oxford: Oxford University Press, 1995), 62; David Hounshell, "Automation, Transfer Machinery and Mass Production in the US Automobile Industry in the Post World War II Era," *Enterprise and Society*, no. 1 (2000): 100.

75 Karel Kaplan, *Sovětští poradci v Československu.* 1949-1956 (Praha: Ústav pro Soudobé Dějiny AV ČR, 1993); Fabio Bettanin, *Stalin e L'Europa. La formazione dell'impero esterno sovietico (1941-1953)* (Roma: Carocci, 2006), 309.
76 Kaplan, *Sovětští poradci*, 7.
77 Ibid., 7-44.
78 Bettanin, *Stalin e L'Europa. La formazione dell'impero esterno sovietico (1941-1953)*, 308. He stresses Gottwald's responsibility in the trial, and the fact that Moscow did not act as a "deus ex machina" and also that the Czechoslovaks themselves informed Moscow of purges that had not been carried out.
79 Kaplan, *Komunistický režim a politické procesy*, 136-42. The trials against those responsible for the management of the economy between 1950 and 1955 involved 50,000 persons; they were mainly middle-class people, but also workers and peasants.
80 *Návrh formulace předběžného československého stanoviska v otázce koordinace plánů výroby automobilů a motocyklů*, September 19, 1949, in SÚA, HR ÚV KSČ, 43, 306. *Rámcové předpoklady pro výrobu vozů Škoda 1200 a T 600, v Automobilových závodech n.p. Ml- Boleslav, v roce 1951 a 1952, Ml. Boleslav*, November 14, 1950, in SÚA, Čzal, b. 86.
81 Kaplan, *Sovětští poradci v Československu.* 1949-1956, 46 footnote no. 61: according to Kaplan, the influence of the Soviet advisors on Czechoslovak economic authorities was enormous between 1950 and 1953.
82 Škoda 1101 (Tudor) was produced in Mladá Boleslav from 1946 to 1952; Škoda 1102 was produced from 1948 to 1952; Škoda 1200 four doors was produced from 1952 to 1956: this was the first Škoda automobile with all-metal bodywork, and it was also produced as a station wagon and multipurpose vehicle. For the technical details: Cedrych and Nachtmann, *Škoda. Auta známá i neznámá* and also: Jan Králík, *Utajené projekty Škoda. Strhující příběh konstruktéra Oldřicha Meduny* (Praha: Grada, 2007).
83 *Zpráva Čsl. delegáta v býro Rady hospodářské vzájemné pomoci*, February 1949 in SÚA, HR ÚV KSČ, 43, 306.
84 *Rámcové předpoklady pro výrobu vozů Škoda 1200 a T 600.*
85 *SUP III a Ministerstvo průmyslu*, November 20, 1950, in SÚA, Čzal, b. 86. With the letter of December 8, 1950, the Minister of Industry indicated to the general manager of ČZAL, who in turn communicated the transfer to AZNP on December 12, 1950.
86 *Rámcové předpoklady pro výrobu vozů Škoda 1200 a T 600, v Automobilových závodech n.p. Ml- Boleslav*, and *Kontrola plnění plánu, rozbory, 1951-1953*, March 25, 1954, and *Struční rozbory změn výrobního programu od r. 1946 do r. 1953*, signed by Václav Medek in Aša, AZNP, b. 27. However, the Mladá Boleslav plant produced only 2,100 Tatras 600 between 1951 and 1952 and only 6410 Tatra's 805 from 1953 to 1955.
87 *Předběžný návrh reorganisace výroby osobních automobilů nového typu v Automobilových závodech, n.p. Mladá Boleslav*, June 16, 1950, in SÚA, Čzal, b. 86.
88 Aša, AZNP, b. 12, August 14, 1952, *Pasport o závodech s větší výrobní plochou než 2.000 m2 a o všech závodech, kde za okupace byla zbrojní výroba.*
89 SÚA, Čzal, b. 86, November 14, 1950, *Rámcové předpoklady pro výrobu vozů Škoda 1200 a T 600.*
90 Ibid.
91 Ibid.
92 Ibid.
93 Ibid.
94 SÚA, Čzal, b. 86, June 18, 1951.
95 Aša, AZNP, b. 10, February 1, 1951, *Smlouva o předání a převzetí výroby Tatra 600*; AZNP, 10, 1954, *Převedení výroby Tatraplanů do AZNP*; *Historie výroby os. vozů 1101-1102*; other surprising information is found in the correspondence between eng. Josef Gryc and Vladimír Matouš and eng. Jiří Hanzelka and Miroslav Zikmund of November 27, 1951, December 8, 1951, and December 29, 1951. Hanzelka and Zikmund, designers of the Tatra, declared that they did not understand what had happened with the S1200 and seemed to expect a reply from Mladá Boleslav. Contrary to the suggestions of the Soviets, they express the hope that the construction of the 8-cylinder Tatra will be completed: "If we used the body prototype that Kopřivnice sent to Montecarlo and that participated at Interlaken, brown and beige with two doors, it could still be a perfect car in ten years' time, and the world would look on in wonder, and this is also true for the 1200 of Mladá Boleslav: both models, a bit prettied up in the finishings, would show the world what we know how to do in Czechoslovakia. Forward motorization!" Aša, AZNP, b. 5, *Změny výrobních programů*, 1954.

Chapter 4

1 Ota Šik, *Quale comunismo?* (Bari: Laterza, 1977) (or. *Das kommunistische Mastsystem*, Hamburg: 1976); cf. Joseph Alois Schumpeter, *Capitalismo, socialismo e democrazia* (Milano: Etas, 2001), 409-10 (or. *Capitalism, Socialism and Democracy*, New York: Harper and Brothers, 1942).

2 Karel Kaplan, "La crisi cecoslovacca," *Annali della Fondazione Feltrinelli* (Milano: Feltrinelli, 1982), 267-326 (published in Czech as: *Sociální souvislosti krizí komunistického režimu v letech 1953-1957 a 1968-1975*, Praha: Ústav pro soudobé dejiny, AV ČR, 1993); Mark Kramer, "The Early Post-Stalin Succession Struggle and Upheavals in East Central Europe: Internal-External Linkages in Soviet Policy Making (Part 1)," *Journal of Cold War Studies* 1, no. 1 (Winter, 1999): 3-55.

3 Jiří Valenta, "Czechoslovakia a 'Proletariat Embourgeoisé,' in *Blue-Collar Workers in Eastern Europe*, eds. Jan Triska and Charles Gati (London: George Allen, 1981), 213; Otto Ulč, "Plzeň: the Unknown Revolt," *Problems of Communism* 14, no. 3 (May-June, 1965): 46-49.

4 Nelson, *Taylor e la rivoluzione manageriale*, 153-65.

5 Yves Cohen, "Les Chefs, une question pour l'histoire du XXe siècle," *Cités* 6 (2000): 4; Nolan, *Visions of Modernity: American Business and the Modernization of Germany*, 180; Bigazzi, *Modelli e pratiche organizzative*, 897-962. As regards the "relationship between the profoundly democratic character of Czechoslovak political culture and Czechoslovak Stalinism" in relation to the trauma of Munich, see Rupnik, *The Roots of Czech Stalinism*, 302-20; and Laudiero, "Peculiarità della cultura politica ceca," in Laudiero, *Api e cinghiali. Culture politiche dell'Europa centrale: il caso ceco*, 183-210.

6 Bloomfield, *Passive Revolution: Politics and the Czechoslovak Working Class 1945-1948*, 91-105; Dorothy Douglas, *Transitional Economic Systems: The Polish Czech Example* (London: Routledge, 1953), 176-290.

7 Series of minutes concerning the *Samospráva*; see in particular those of June 1947; December 21, 1947; March 12, 1947; March 12, 1948; in Aša, AZNP/P, b. 12. In this connection the minutes of June 19, 1948, on the introduction of the *System Bat'a* are interesting. Cf. Václav Tágl's report on Bat'a: "The visit took place from 25-27 June at Zlín. It was authorized and organized by the factory council and the factory manager." The author concluded: "I think it will be necessary to introduce this system at Mladá Boleslav, too, simply by giving the order, without any discussion or debate. Workers who aren't afraid of work and really want to work have nothing to fear from this system. I feel that in this way we will automatically solve the problem of the reduction of office workers and the problem of timekeepers. The authority of the managers will be restored, and the factory will prosper once again. I think that trust in the conscience of employees and in voluntary training is doomed to failure. When we see the economic results of Bat'a (in building factories, in employees' dwellings, in cultural activities, in the research institute, in theatre and culture) or the people's well-being, we have to admit that the advantages of this system outweigh the supposed 'apparent' constriction to a degrading form of work. The best publicity for this system would be mass outings."

8 Aša, AZNP, b. 32.

9 Valenta, *Czechoslovakia a "proletariat embourgeoisé,"* 213; Rupnik, *The Roots of Czech Stalinism*, 310-11.

10 Merkle, *Management and Ideology: The Legacy of the International Scientific Management Movement*, 103; Anfossi, *Prospettive sociologiche sull'organizzazione aziendale*, 43-44.

11 Cf. Vladimir I. Lenin, *Opere scelte* (Moscow: Progress Publishers, 1971): especially Vladimir I. Lenin, "Stato e rivoluzione. La dottrina marxista dello stato e i compiti del proletariato nella rivoluzione," 268-359 (published in US as: *State and Revolution: Marxist Teaching about the Theory of the State and the Tasks of the Proletariat in the Revolution*, Westport, Conn: Greenwood Press, 1978) and "I compiti immediati del potere sovietico," *Opere scelte*, 412-43 ("The Immediate Tasks of the Soviet Government," in *Collected Works*, 4th English Edition, vol. 27, 235-77, Moscow: Progress Publishers, 1972. First published in *Pravda* in April 1918).

12 Lenin, *I compiti immediati*, 427.

13 Ibid., 428; Maier, *Between Taylorism and Technocracy*, 51.

14 Lenin, *Stato e rivoluzione*, 303, 309, 332.

15 Lenin, *I compiti immediati*, 427.

16 Ibid., 429.

17 Lenin, *Stato e rivoluzione*, 304. Cf. Max Weber, *Sul Socialismo reale* (Firenze: Savelli, 1979), 28-39 (or. *Der Sozialismus. Rade zur allgemeinen Orientierung von oesterreichischen offizieren in Wien 1918*, Gesammelte Aufsaete zur Soziologie und Sozialpolitik, 1924).

18 Nicolas Werth, *Storia dell'Unione Sovietica. Dall'impero russo alla Comunità degli stati indipendenti. 1900-1991*, trans. M.R. Baldi (Bologna: Il Mulino, 1994), 147-98; see also Sheila Fitzpatrick, *The Russian Revolution* (Oxford: Oxford University Press, 1994), 81; Edward H. Carr, *La rivoluzione russa. Da Lenin a Stalin (1917-1929)*, trans. Franco Salvatorelli (Torino: Einaudi, 1980) (or. *The Russian Revolution from Lenin to Stalin 1917-1929*, New York: Free Press, 1979), 24-34; Andrea Graziosi, *Stato e Industria in Unione Sovietica (1917-1953)* (Napoli: Edizioni Scientifiche Italiane, 1994).

19 Marc Beissinger, *Scientific Management, Social Discipline, and Soviet Power* (London: I.B. Tauris & Co., 1988), 32.

20 Salsano, *Ingegneri e politici. Dalla razionalizzazione alla "rivoluzione manageriale"*, 95-106.

21 Beissinger, *Scientific Management, Social Discipline, and Soviet Power*, 71, 161 and 25-27; Merkle, *Management and Ideology: The Legacy of the International Scientific Management Movement*, 123.

22 In January 1921 the first Soviet congress on NOT was held, and these positions were held by Ossip Hermanski. In 1922 the Moscow Central Institute of Labor (TsIT) was set up, and these projects were followed by the People's Commissariat and by the Komsomol [*Kommunisticeskij Sojuz Molodezi- Soviet Youth Organization*], on which cf. Antonio Martelli, *Teorie e ideologie del management. Profilo storico delle dottrine manageriali (1770-1970)* (Milan: ETAS libri, 1979), 129. Positions strongly opposed to Taylor's system had been held in the early 1920s in the trade unions by the Mensheviks, who refused the system in the name of past struggles, as well as by leftist Bolsheviks. Beissinger, *Scientific Management, Social Discipline, and Soviet Power*, 145.

23 Loren Graham, *The Ghost of Executed Engineer: Technology and the Fall of the Soviet Union* (Cambridge: Harvard University Press, 1993), 39; see also Kendall E. Bailes, *Technology and Society under Lenin and Stalin: Origins of the Soviet Technical Intelligentsia 1917-1941* (Princeton: Princeton University Press, 1978), 55; Nick Lampert, *The Technical Intelligentsia and the Soviet State: A Study on Soviet Managers and Technicians 1928-1935* (London: Macmillan, 1979). Indeed, only 2 percent of Soviet engineers were party members in 1927.

24 Carr, *La rivoluzione russa. Da Lenin a Stalin (1917-1929)*, 186-97. In 1929 Nikolaĭ I. Bukharin's *Notes of an Economist*, supporting the "snail's pace" in Soviet industrialization, was condemned. Trotsky was expelled from the Soviet Union and took refuge in Istanbul. See Nikolaĭ I. Bukharin's "Notes of an Economist," in *Selected Writings on the State and on the Transition to Socialism*, ed. Richard B. Day (New York: M.E. Sharpe Inc., 1982).

25 Carr, *La rivoluzione russa. Da Lenin a Stalin (1917-1929)*, 159; Fitzpatrick, *The Russian Revolution*, 120; Moshe Lewin, *Storia sociale dello stalinismo* (Turin: Einaudi, 1985), 272 (or. Moshe Lewin, *The Making of the Soviet System: Essays in the Social History of Interwar Russia*, London: Methuen, 1985).

26 Werth, *Storia dell'Unione Sovietica. Dall'impero russo alla Comunità degli stati indipendenti. 1900-1991*, 274-75; Beissinger, *Scientific Management, Social Discipline, and Soviet Power*, 120; Graham, *The Ghost of Executed Engineer: Technology and the Fall of the Soviet Union*, 73.

27 Lewis H. Siegelbaum, *Stakhanovism and the Politics of Productivity in the USSR, 1935-1941* (Cambridge: Cambridge University Press, 1988).

28 Lewin, *Storia sociale dello stalinismo*, 39.

29 Weber, *Sul socialismo reale.*

30 Lewin, *Storia sociale dello stalinismo*, 272-94. On the relationship between Stakanovism and Taylorism compare, among others: Merkle, *Management and Ideology*, 128; Lewis H. Siegelbaum, *Stakhanovism and the Politics of Productivity in the USSR, 1935-1941* and idem, "Master of the Shop Floor: Foremen and Soviet Industrialization," in *Stalinism: Its Nature and Aftermath, eds.* Nicholas Lampert and Gabor Rittersporn (London: MacMillan, 1992), 127-56.

31 Aša, AZNP, b. 5, file: *Monthly reports of the Mladá Boleslav factory for the year 1951*; Aša, AZNP, b. 7, file: *Zápisy. Konference závodních ředitelů 1951, 1952, 1953.*

32 On the basis of agreements among the allied forces at the Postdam Conference in July-August 1945, with the consensus of a large majority of the country's political class, the thirteen decrees known as the Decrees of Beneš (President of the Republic at the time) were drawn up. These decrees sanctioned the Czechoslovak determination to create a strong Slavic nation state composed of Czechs and Slovaks. The consequence was that between 1945 and 1946, three million German residents of Czechoslovakia were expelled or left Czechoslovakia, dropping, according to Krejčí's data, from 22 percent of the population in 1930 to 1 percent in 1950. There was also an exodus of half a million Hungarians, who returned to Hungary from Slovakia. To see the significance of these figures we need only consider that in 1937 the Czechoslovak population was roughly fifteen million, a level that would only be

reached again in 1980. Teichová, *The Czechoslovak Economy*, 94-98; and Jaroslav Krejčí, *Social Change and Stratification in Postwar Czechoslovakia* (London: Macmillan, 1972), 8.

33 The expulsion of German workers who according to Myant represented 30 to 40 percent of the workforce and were concentrated in some zones of Czechoslovakia, at the border and in the northwest.

34 On the question of female workers, Alice Teichová relates that between 1950 and 1972, 1,602,000 workers entered Czechoslovak factories, of whom 1,258,000 were women, equal to 78 percent of the increased workforce. Aša, AZNP, b. 5, "Ženy které vyrábějí motorová vozidla," in *Vlasta. List Výboru čs žen*, vol. 43 (1954).

35 Aša, AZNP/P, b. USSR, file: *Rozbor otázek pro SSSR*, question no. 27, 11. The percentage of women working at small presses was 90 percent, and they were being considered for the large presses as well; the ratio between large and small presses was 8:1, according to data from 1948.

36 Ibid.

37 Ibid., questions 69-70, 22-23.

38 Ibid.

39 Aša, AZNP, b. USSR, question no. 27, 11.

40 Granick, *Il dirigente Sovietico. Studio sull'uomo di organizzazione nell'industria russa* (Milan: Edizioni di Comunità, 1962), 62 (or. *The Red Executive: A Study of the Organization Man in Russian Industry*, Garden City, N.Y.: Doubleday, 1960).

41 *Rozbor otázek pro SSSR.*, question no 29, 13.

42 Ibid., question no 13, 3.

43 Ibid.

44 Lewin, *Storia sociale dello stalinismo*, 288.

45 *Rozbor otázek pro SSSR.*, question no. 83, 29-30.

46 Ibid., question no. 83, 29.

47 Ibid., question no. 83, 34.

48 Ibid., question no. 81, 25.

49 Ibid., question no. 29, 12.

50 This recalls the description of Hungarian shops in the 1970s in Miklos Haraszti, *A cottimo. Operaio in un paese socialista* (Milan: Feltrinelli, 1975) (or. *Worker in a Worker's State: Piecework in Hungary*, New York: Universe Books, 1975); see also Burawoy and Lukas, *The Radiant Past*, 4-6.

51 Kaplan, *La crisi cecoslovacca*, 271.

52 Yves Cohen, "Administration, politique, et techniques: réflexions sur la materialité des pratiques administratives dans la Russie stalinienne (1922-1940)," *Cahiers du monde russe* 44, no. 2/3 (Apr.- Sept. 2003): 269-307.

53 Gerald A. McDermott confirms these positions: *Embedded Politics: Industrial Networks and Institutional Change in Postcommunism* (Ann Arbor: Michigan University Press, 2002), 32-33.

54 Aša, AZNP, b. 1, file: *Usnesení předsednictva ÚV KSČ a vlády ze dne 27 dubna 1953 o opatřeních k upevnění technologické kázně ke snížení počtu zmetků ve výrobě a ke zvýžení kvality výrobků.*

55 Ibid.

56 Ibid.

57 Ibid.

58 Kaplan, *Československo v RVHP*, 55-75.

59 *Oborová konference pro osobní automobily konaná dne 30-31 března 1956. Důverné* intervention on normalization, standardization, patents, and innovation, signed by Z. Mauermann, director of standardization department in AZNP, Mladá Boleslav, in Aša, AZNP, b. 8.

60 Cf. "Likvidace soukromých živnostníků 1948-1950," in Zdeněk Kárník and Jan Měchýř, *K novověkým sociálním dějinám Českých zemí IV. Zvraty a převraty 1939-1992* (Praha: Karolinum, 2001). See also the work edited by Zdeněk Kárník and Michal Kopeček, *Bolševismus, komunismus a radikální socialismus v Československu* (Praha: Ústav pro soudobé dějiny AV ČR, Dokořán, 2003), vol. 1. See also the studies of Pavel Machonin: Pavel Machonin, *Československá společnost* (Bratislava: Epocha, 1969); Pavel Machonin and Jaroslav Krejčí, *Czechoslovakia 1918-1992* (Basingstoke: Macmillan, 1996).

61 Aša, AZNP, b. 1, file: *Statut oddělení technické kontroly strojír. podniku* and series of resolutions of the Presidency of the Central Committee of the KSČ for the years 1952 and 1953 esp. *Organizační změny 2. září 1952* and series of *Organisační předpis podnikového ředitele*, 1952.

62 Aša, AZNP, b. 11, July 1953, file: *Zavádění dispečerské služby v automobilových závodech n.p. Mladá Boleslav*, signed by Bedřich Nosek (as a manual for training MVS middle managers), introduction of

the *dispečerská služba* is a "Christmas gift", the result of a director's visit to the USSR in June 1953.
63 Cohen, *Administration, politique, et techniques: réflexions sur la matérialité des pratiques administratives dans la Russie stalinienne (1922-1940)*, 269.
64 Beissinger, *Scientific Management, Social Discipline, and Soviet Power*, 145.
65 Ibid.
66 Cohen, *Administration, politique, et techniques*, 279.
67 Ibid.
68 Cohen, *Administration, politique, et techniques*, 281.
69 Aša, AZNP, b. 11, file: *Training Middle Managers. Document of June 24, 1953.*
70 Ibid.: they were former shop assistants who had worked on executing the plan.
71 Stanislav Císař, Director of the enterprise from October 1, 1950, appointed by the Director of Čzal, Horák; cfr. Aša, AZNP, b. 29, file: *Document in Kadrové věci*, March 13, 1956.
72 Aša, AZNP, b. 11, file: *Organisační předpis podnikového ředitele, 6\52,* October 13, 1952.
73 Aša, AZNP, b. 11, file: *Training of cadres*, June 24, 1953. The new model was the Spartak-Octavia Š440.
74 Aša, AZNP, b. 11, file: *Chozrasčot-zase něco nového?* March 26, 1953 and *Zavádění dispečerské služby v Automobilových závodech nár. podnik. Mladá Boleslav*, author Bohuslav Nosek and *Projekty-Organisace Řízení*, 1951-1954 and *Vnitrozávodní chozrasčot*: the perplexities mainly concerned the continual organizational changes and the introduction of new paperwork and administration that "weighed heavily on manufacturing shops". The writer defended himself by saying that the new organization was suited to the socialist emulation and the planned economy that had saved Czechoslovak workers from the scourge of unemployment. It allowed clear responsibilities to be established. Bonuses were no longer only for the foreman, as with the capitalist Škoda, but were shared among all the workers; everyone shared the advantages of improved productivity, and savings were made with the product until the end. Every worker became *hospodář*. Moreover, this was not merely a passive adjustment: value was given to what already existed at Škoda; for example, accounting machines were used, rather than returning to doing bookkeeping by hand.
75 Aša, AZNP, b. 11, *Chozrasčot-zase něco nového?* March 26, 1953.
76 Kaplan, *La crisi cecoslovacca 1953-1956*, 267-326.
77 Jiři Pernes, *Snahy o překonaní politicko-hospodářské krize v Československu v roce 1953*, Brno, Ústav pro soudobé dějiny, AV ČR, 2000, 9: controversial because at the same time the presence of two markets favored workers in heavy industry and office workers faithful to the regime, as well as working women. According to Pernes, in the end the double market had more disadvantages than advantages, and they decided to abolish it. See also Jiří Pernes, "Dělnické demonstace v Brně v roce 1951," *Soudobé dějiny*, 1 (1996): 23-41, on the strikes caused by the abolition of the *vánoční příspěvek* (Christmas bonus).
78 Kaplan, *La crisi cecoslovacca*, 267. The demonstrations took place mainly in Třinec, Ostrava, Stalingrad ČKD, Pardubice, Kladno and Vimperk.
79 Ulč, *Plzeň: the Unknown Revolt*, 46-49. Unfortunately, the report, which is extremely rhetorical, of a former judge of the district court who escaped to the U.S. some years after the fact is the only direct source that recounts the event in detail. Recently, the Czech press has gathered other statements made by witnesses from the Archives of the CP and some protagonists, but a systematic study of the question is still lacking. Cf. Karel Bartošek, "Stávkujicí dělník v práci, roku 1953," *Listy*, 17, no. 2 (1987); Kramer, *The Early Post-Stalin Succession Struggle and Upheavals in East Central Europe*, 3-55: this work places strong emphasis on the relationship between the instability of the area under Soviet influence and Stalin's death and on the importance of looking closely at internal politics to understand the changes in the system. The events of Plzeň are also seen in relation to the influence they had on Soviet politics. Initially, the Soviets were frightened by the size of the demonstrations and urged reforms, despite local leaders' resistance. Later, however, with Nikita Chruschev's victory, another line prevailed. Cf. Myant, *The Czechoslovak Economy 1948-1988*, 64: the strikes involved 129 factories and 32,000 workers.
80 There are very few recent studies of the Czechoslovak "working class"; in particular, see: Peter Heumos, *"Vyhrňme si rukávy, než se kola zastaví!" Dělníci a státní socialismus v Československu, 1945-1968* (Praha: Ústav pro soudobé dějiny AV ČR, 2006); and Jiří Maňák, "Proměna dělnické strany v organizaci moci. Problematika dělnického charakteru KSČ v letech 1945-1953 ve světle stranické statistiky," in *Bolševismus, komunismus a radikální socialismus v Československu*, eds. Kárník and Kopeček, 157-99.

81 Pernes, *Snahy o překonaní politicko-hospodářské krize v Československu v roce 1953.*
82 Dalibor Státník, *Sankční pracovní právo v padesátých letech. Vládní nařízení o opatřeních proti fluktuaci a absenci č.52\1953 Sb* (Praha: Ústav pro soudobé dějiny AV ČR, 1999). The Central Union Council, URO, charged with keeping up "labor morale", participated in drawing up the decree. The measure was considered the final attempt to manage the chaos created by the Five-Year Plan using Stalinist methods. According to the author, the opposition to the decree is proof that by then, human reserves were exhausted, and industry had to follow new paths to grow.
83 Aša, AZNP, b. 15, file: *Příkaz ministra všeobecného strojírenství číslo 192\53*, June 4, 1953.
84 Aša, AZNP, b. 15, file: *Mzdová opatření v důsledku usnesení vlády a strany o provedení peněžní reformy a zrušení lístků.*
85 Ibid.
86 Ibid.
87 Ibid., *Zajištění výsledků peněžní reformy a zrušení lístkového systému*, see the ministerial text of July 13, 1953: in Aša, AZNP, b. 15, which reads as follows: "1. They must motivate [the workers] to an ever more profound application of the principle of personal responsibility: it is always necessary to immediately give every task to an employee who will execute it or will guarantee and be responsible for its execution. 2. They must motivate them to carry out tasks through live, effective and concrete direction of work, without creating useless intermediate passages. 3. Diligently assure and firmly declare discipline in every part of the state and economic apparatus and with every worker. 4. Collaborate and assure timely, effective and systematic control of tasks on a daily basis; effect controls both on the preparation of tasks and their execution. 5. Increase the political and professional quality of the employees of the state and economic apparatus. 6. Allocate employees correctly according to their qualities and correctly organize the distribution of tasks to employees. 7. Insist on giving importance to growth in the level of the employees of the state and economic apparatus based on criticism and self-criticism."
88 Mark Kramer, "The Early Post-Stalin Succession Struggle (part three)," *Journal of Cold War Studies*, 1, no. 3 (Autumn, 1999): 66.
89 Pernes, *Snahy o překonaní politicko-hospodářské krize v Československu v roce 1953*, 16.
90 Aša, AZNP, b. 1, Stanislav Císař, *Spolupráce s MS ministr Poláček, Kritické připomínky k práci ministerstva strojírenství*, November 20, 1953.
91 Ibid.
92 McDermott, *Embedded Politics*, 30-33; A. Teichová, *The Halting Pace to Scale and Scope*, 447.
93 *Kritické připomínky k práci ministerstva strojírenství,*
94 Josip Stalin, *Questioni del Leninismo* (Roma: Ed.Unità, 1945).
95 Kaplan, *The Overcoming of the Regime Crisis After Stalin's Death in Czechoslovakia, Poland and Hungary*, 291.

Chapter 5

1 Myant, *The Czechoslovak Economy*, 66-70 provides data and clearly describes the economic background of the apparent success of the Czechoslovak economy in 1953-56 and the economic benefits of the New Course; Francesco Leoncini, "Intellettuali e operai nel '56 ungherese," in Leoncini, *L'Europa Centrale. Conflittualità e progetto*, 191-99; Francesco Leoncini, ed., *Che cosa fu la "Primavera di Praga". Idee e progetti di una riforma politica e sociale* (Manduria: Lacaita, 1989). On the workers' councils, see also: Peter Heumos, *Vyhrňme si rukávy než se kola zastaví*, 65 and Mark Pittaway, "Introduction: Workers and Socialist States in Postwar Central and Eastern Europe," *International Labor and Working Class History* 68 (2005): 1-8.
2 Rupnik, *The Roots of Czech Stalinism*, 302; Laudiero, *Peculiarità della cultura politica ceca*, 206-07.
3 Kaplan, *La crisi cecoslovacca*, 306.
4 Jiří Kosta, "Storia e contenuti della riforma economica cecoslovacca negli anni 1965-69," in *Che cosa fu la "Primavera di Praga"*, ed. Leoncini, 41-67; Concerning the workers' councils and the different positions among "progressive" forces, Karel Kovanda, "I consigli operai in Cecoslovacchia (1968-1969)," in *Che cosa fu la "Primavera di Praga"*, 69-90; cf. Boffito and Foa, *La crisi del modello sovietico*

in Cecoslovacchia, 170-211, 219-23, 233-37 and Mark Kramer, "New Sources on the 1968 Soviet Invasion of Czechoslovakia (Part One and Part Two)," *Cold War International History Project Bulletin*, Fall 1992, no. 2, Woodrow Wilson International Center for Scholars, Washington DC.

5 In 1962 Novotný appointed Ota Šik, director of the Institute of Economics of the Academy of Sciences, as head of the commission of economic experts; the directives for the Fourth Five-Year Plan (1966-70) elaborated by the same commission contain the basic principles of economic reform, introduction of the market mechanism, planning autonomy for enterprises in every economic sector, business profitability accounting, state ownership of the means of production. See: Myant, *The Czechoslovak Economy*, 220 and 98-110; Brus, "1950 to 1953: The Peak of Stalinism," 130; Victor Nee and David Stark, eds., *Remaking the Economic Institutions of Socialism, China and Eastern Europe* (Stanford: Stanford University Press, 1989).

6 Fréderic Fléron, *Technology and Communist Culture, the Socio-Cultural Impact of Technology Under Socialism* (New York: Praeger, 1976), 5.

7 Muriel Blaive, *Une déstalinisation manquée. Tchècoslovaquie 1956*, préface de Krzysztof Pomian, Editions Complexe, 2005, and Jaromír Navrátil, ed., *The Prague Spring, 1968: A National Security Archive Documents Reader* (Budapest: Central European University Press, 1998).

8 Blaive, *Une déstalinisation manquée. Tchècoslovaquie 1956*, 17.

9 Ibid. 175.

10 Myant, *The Czechoslovak Economy*, 74-75.

11 Raymond Stokes, "In Search of the Socialist Artefact: Technology and Ideology in East Germany, 1945-1962," *German History* 15 (1997): 221-39, spec. 225. See also Raymond Stokes, *Constructing Socialism: Technology and Change in East Germany 1945-1990* (Baltimore: The Johns Hopkins University Press, 2000), 112-13; and David Crowley and Susan Reid, "Style and Socialism: Modernity and Material Culture in Post-War Eastern Europe," in *Style and Socialism: Modernity and Material Culture in Post-War Eastern Europe*, eds. David Crowley and Susan Reid (Oxford: Berg, 2000), 1-24; in the same volume, on pages 65-80 there is another essay by Raymond Stokes, "Plastics and the New Society: the German Democratic Republic in the 1950s and the 1960s". See also Siegelbaum, *Cars for Comrades: The Life of the Soviet Automobile*, cit.; Jonathan R. Zatlin, "The Vehicle of Desire: The Trabant, the Wartburg, and the End of the GDR," *German History* 15 (1997): 358-80.

12 Miroslav Rosický, *Organizace výrobní základny a plánovité řízení ekonomiky* (Prague: Akademie, 1983), 212.

13 *Oborová konference*, March 30-31, 1956 in Aša, AZNP, b. 8.

14 One needs only to think of the American exhibition in Moscow in 1959, and obviously of the launch of Sputnik in 1957; on these topics, see Paul R. Josephson, "Projects of the Century in Soviet History: Large Scales Technologies from Lenin to Gorbachev," *Technology and Culture* 34, no. 3 (1995): 519-59 and the very recent volume edited by Ruth Oldenziel and Karin Zachmann, *Americanization, Technology and European Users* (Cambridge: The MIT Press, 2009). In their introduction, the authors write on the conflictual relationship with individual consumption: "No wonder that the American way of defining *technological advancement* in consumer terms in their public relations exasperated the Soviets. For the Soviets, the emphasis on individual consumer goods, moreover, was a moot rhetorical point. Soviet leaders were dedicated to technological systems that would be accessible to and affordable for all citizens [...]. During the late 1950s, Soviet leaders may have felt pressured into allowing some private consumption to shore up their authority, but in terms of economic policy the Soviets focused their efforts first and foremost on rebuilding production capacity than encouraging individual consumption", 7; a similar position is expressed by Lewis Siegelbaum in his work on the history of the Soviet automobile, where he writes of "the Faustian Agreement" between the Party and the population regarding individual motorization: Siegelbaum, *Cars for Comrades: The Life of the Soviet Automobile*. Along the same lines of interpretation, see André Steiner, *Dissolution of the Dictatorship Over Needs? Consumer Behaviour and Economic Reform in East Germany in the 1960s*; and Ida Merkel, "Consumer Culture in the GDR" or "The Struggle for Antimodernity was Lost on the Battleground of Consumer Culture", both in *Getting and Spending: European and American Consumer Societies in the Twentieth Century*, eds. Susan Strasser, Charles McGovern and Matthias Judt (Washington, DC: Publications of the German Historical Institute, 1998), 167-85, 281-301. See also David Crowley and Susan Reid, eds., *Socialist Spaces: Sites of Everyday Life in the Eastern Bloc* (Oxford: Berg, 2002), 123-40.

15 Stokes, *In Search of the Socialist Artefact: Technology and Ideology in East Germany, 1945-1962*, 221-39.

16 Tim Edensor, "Automobility and National Identity: Representation, Geography and Driving Practice,"

Theory, Culture and Society 21/4-5 (2004): 101-20 and John Urry, "The 'System' of Automobility," *Theory, Culture and Society* 21/4-5 (2004): 25-39.

17 Peter Heumos, *"Vyhrňme si rukávy než se kola zastaví". Dělníci a státní socialismus v Československu, 1945-1968,* 41.

18 Jan Králík, *V soukolí okřídleného šípu,* and František Žalud, *Přežili jsme.*

19 Françoise Mayer, *Les Tchèques et leur Communisme. Mémoire et identités politiques* (Paris: Editions de l'Ecole des Hautes Etudes en Sciences Sociales, 2004), spec. 225; and the work by Ladislav Holy, *The Little Czech and the Great Czech Nation: National Identity and the Post-Communist Social Transformation* (Cambridge: Cambridge University Press, 1996), 72.

20 Úvod in SÚA, ser. Ministerstvo automobilového průmyslu a zemědělských strojů (henceforth MAP) and ser. Ministerstvo všeobecného strojírenství (henceforth Mvs).

21 *Oborová konference pro osobní automobily,* March 30-31, 1956 in Aša, AZNP, b. 8.

22 "Naše cesta k lidovému vozu," *Svět motorů* 164 (1954): 4-5, continuing in issue 165 (1954): 38-39. In the preceding issue 161-62 (1953) there was already an article on this topic: "Malý automobil-velký problém," 800.

23 Stokes, *Constructing Socialism: Technology and Change in East Germany,* 112-13.

24 *Návrh perspektivního plánu oboru osobní automobily od roku 1956 do roku 1975* in SÚA, MAP, b. 200.

25 "Spartak-mezityp lidového automobilu," *Svět motorů* 189 (1954), 756-57; "Ještě něco o Spartaku," *Svět motorů* 190 (1954): 53; "Deset let vývoje a výroby automobilů," *Svět motorů* 198 (1954): 298-99; "Více, lépe a levněji," *Svět motorů* 204 (1955): 1; "Lépe využívat zahraničních technických poznatků a zintensivnit propagaci nové techniky," *Svět motorů* 209 (1955): 1.

26 "Prodej osobních automobilů v roce 1957," *Svět motorů* 240 (1956): 823; "Prodej osobních automobilů obyvatelskvu v roce 1956," *Svět motorů* 218 (1956): 122; "Kdy a jak si budeme moci koupit automobil," *Svět motorů* 244 (1957): 107-08.

27 "Motoristé před volbami," *Svět motorů* 250 (1957): 1; see also "V Italii a u nás, na pomoc předvolebním besedám," *Svět motorů* 195 (1956): 742.

28 "K některým otázkám daně z motorových vozidel," *Svět motorů* 266 (1957): 1; Karel Šimon, "Strana a vláda pečují o rozvoj motorismu," *Svět motorů* 261 (1957): 1; Václav Jirout, "Více, lépe a levněji," *Svět motorů* 262 (1957).

29 *Návrh perspektivního plánu oboru osobní automobily od roku 1956 do roku 1975* in SÚA, MAP, b. 200.

30 *III etapa plánu rozvoje a specialisace pro osobní automobily 1955-1960* in Aša, AZNP, b. 26 and 34. For technical and construction details of the models, cf. Kožíšek, Králík, *L&K-Škoda;* for a full description of the models produced and production methods, cf. Cedrých and Nachtmann, *Škoda. Auta známá i neznámá. Prototypy i seriové automobily vyráběné od roku 1934,* 141-55.

31 SÚA, MAP, b. 94, 7. There were twenty-five projects for interventions for reconstruction of plants and purchase of machinery, with an investment of over 25 million crowns, of which 2.36 million was for construction.

32 SÚA, MAP, b. 94, 6.

33 *Rozvoj automobilového průmyslu v ČSR,* 1970 in Aša, AZNP, b. 34 and *Podnikatelský záměr na výrobu osob. aut. v ČSSR do roku 1980, April 1968* in Aša, AZNP, b. 65: available data show that production reached 77,000 units in 1965, and 125,000 units in 1968, with a growth of 95 percent between 1962 and 1968. This is even more noteworthy if compared with 1957, when the Mladá Boleslav lines produced only 35,000 automobiles. Still, these numbers were very small compared with foreign production figures (in 1968 Democratic Germany produced 2.9 million pieces, and Italy 1.5 million). For a comparison with the Italian case, see: D. Bigazzi, *La grande fabbrica,* 179, and with the world-wide industry: J. Bardou, *The Automobile Revolution.*

34 In Aša, AZNP, b. 65: according to the same source, the Federal Republic of Germany had increased from one car for every 20 inhabitants to one for every 5.1 in 1967, in Great Britain from one car for every 11 to one for every 5, and in France from one for every 12 to one for every 4.5.

35 *Oborová konference pro osobní automobily,* and Miroslav Vigner, "Vliv účelného uspořádání výrobního pochodu na hospodárnost," *Podniková organisace* 7, 1955, 264-67.

36 Miroslav Vigner (1920-1982) was trained at the Prague Politechnic ČVUT in 1938. He worked at the AZNP from 1947 to 1949 as a designer, and between 1949 and 1950 he worked for the KOVO exporting vehicles. Between 1950 and 1980 he was at AZNP, first as director of the press department and then as development secretary. In 1966 he was appointed secretary to the director, Josef Šimon. He taught at ČVUT, in the mechanical engineering faculty. Králik, *V soukolí okřídléného sípu* 46.

37 *Oborová konference*, intervention of Miloslav Vigner.
38 Ibid.
39 Ibid.
40 Ibid.
41 Ibid.
42 Ibid., intervention of Zdeněk Kejval, 62-88.
43 *Zpráva komise politického byra UV KSČ, pres. K. Poláček, 8-7-1957. Zpráva o plnění usnesení politického byra UV KSČ ze dne 16-4 a 9-10-1957 o opatřeních k zajištění kvality nakladních automobilů Š706 a osobních automobilů Š440*, December 14, 1957, in SÚA, MAP, b. 30; *Zpráva pro sekretariát ministra o organizaci, technologii a efektivnosti výroby, o výzkumu, kvalitě vyrobků, 1955-1958* in SÚA, MAP, b. 94.
44 *Specialisace a kooperace jako jeden z vážných činitelů ovlivňujících dálší rozvoj vyroby ozubených kol a převodových skříní pro automobily, září 1957* in Aša, AZNP, b. 48.
45 *Zpráva o výsledku kontroly uskutečňování usnesení únorového zasedání ústředního výboru komunistické strany Československa o zvýšení efektivnosti národního hospodářství v souvislosti s připravu plánu na rok 1958 podle usnesení vlády č.848.*
46 *Návrh plánu výroby-rok 1956. k 4 závěrce. AZNP MB* in SÚA, MAP, b. 188; *Výstavba automobilových závodů n.p. Mladá Boleslav, 1961-1965* in SÚA, Mvs, b. 186; *Zpráva o stavu zajištění dodávek technologického zařízení, strojů, projekce,a stavebních prací pro centralisovanou výstavbu AZ MB z podniku ministerstva TS, MZO, MV e Mvs*, September 26, 1961, in SÚA, Mvs, b. 190 and *Zvýšení cen u jednoúčelových a upravnených specielních strojů a zařízení*, 1962 in SÚA, Mvs, b. 37; *Zpráva o překročení investičních nákladů na centralisovanou stavbu rekonstrukce a dostavba n. AZMB* 1961, SÚA, Mvs, b. 37. Cf. *Zápisy ze zasedání tech. Ekonomické rady* 1960-1962-1963 in Aša, AZNP, b. 6.
47 *Návrh plánu výroby rok 1956*, SÚA, MAP, b. 188.
48 Miroslav Rosický, "Organization and Management of Industry and Construction in the Czechoslovak Socialist Republic," *Czechoslovak Economic Papers* vol. 12 (Prague: Nakladatelství Československé akademie věd, 1962).
49 Myant, *The Czechoslovak Economy*, 79-85.
50 Teichová, *The Halting Pace to Scope and Scale*, 148
51 Rosický, *Organization and Management of Industry and Construction*, 60-61.
52 *Organisáce jednotlivých výrobné hospodářských jednotek skupině I, silniční doprávní prostředky*, SÚA, MAP, 98, 13. *Prozatímní organisační řád*, 12 July 1948, in Aša, AZNP, 33.
53 Rosický, *Organization and Management of Industry and Construction in the Czechoslovak Socialist Republic*, 42-43.
54 Teichova, *The Halting Pace to Scale and Scope*, 452-53.
55 Boffito and Foa, *La crisi del modello sovietico in Cecoslovacchia*, 35-36.
56 Gerard McDermott, *Embedded Politics: Industrial Networks and Institutional Change in Postcommunism* (Ann Arbor: Michigan University Press, 2002), 78.
57 Gerald McDermott, "Network Restructuring and Firm Creation in East Central Europe: A Public Private Venture," in *The New Entrepreneurs of Europe and Asia*, eds. Victoria Bonnell and Thomas B. Gold (New York: M.E. Scarpe, 2002), 214-55.
58 *Zprávy pro segretariát ministra, O produktivitě práce a mzdách,* in SÚA, MAP, b. 96.
59 *Zpráva o přestavbě soustav dělnických mezd a politickoorganisačním zajištění přestavby v resortu ministerstva automobilového průmyslu a zemědělských strojů,* July 15, 1958, in SÚA, MAP, 188. Rosický, *Organizace výrobní základny a plánovité řízení ekonomiky*, 220.
60 *Oborová konference*, relazione di ing. Rudolf Vykoukal, *Rozbor stavu techniky osobních automobilů v ČSR v porovnání se stavem v zahraničí a se světovou úrovní.*
61 *Oborová konference*, report by Miloslav Vigner, 59.
62 *Zpráva o přestavbě soustav dělnických mezd a politicko organisačním zajištění přestavby v resortu ministerstva automobilového průmyslu a zemědělských strojů,* in SÚA, MAP, b. 188.
63 Ibid.
64 Ibid.
65 Ibid.
66 Jan Pivoňka, "Mistr – organisátor výroby, socialistického soutěžení, rádce a vychovatel," *Podniková organisace* 9 (1955): 350-52.
67 *Zpráva o přestavbě soustav.*

68 Cf. Haraszti, *A cottimo*, on the question of the norms and Burawoy, Lukacs, *The Radiant Past*. In both cases, the description of work methods in the Hungarian factory are very like what comes out in this *cahier des doléances*. For a more methodical historical treatment of the ways norms were determined in Czechoslovakia and of the sense of the reforms, see Jaroslav Jirásek, *Organizační řízení. Vývoj teorie podnikového řízení* (Prague: Institut řizení, 1987), 131-63.
69 *Zápis z aktivu s delegaty Motokovu ze zahraničí zástupci Motokovu se zástupci AZNP*, December 11, 1958 in Aša, AZNP, b. 8a.
70 Obviously, the reasons for the difficulties of Czechoslovak cars on western markets were not all technical in nature; on the contrary, it needs to be said that starting from the late 1950s domestic demand was increasing inside the Soviet-block countries, and obviously the Comecon was also more active. I take up some of these arguments with greater contextualization in Valentina Fava, "The Automobile Industry and CMEA Integration: Evidence from the Czechoslovak Case and Reflections on an Unexpected Failure," *Jahrbuch für Wirtschaftsgeschichte / Economic History Yearbook* 2 (2008): 93-115.
71 Keith Pavitt, "Alcuni fondamenti della teoria della grande impresa innovativa," in *Innovazione, impresa e sviluppo economico*, eds. Renato Giannetti and Pier Angelo Toninelli (Bologna: Mulino, 1991); Nathan Rosenberg, *Dentro la scatola nera: tecnologia e economia* (Bologna: Mulino, 1991) (or *Inside the Black Box. Technology and Economics*, Cambridge: Cambridge University Press, 1982).
72 SÚA, ser. Úřad předsednictva vlády – tajná spisovna (henceforth ÚPV-T), b. 48, September 14, 1973.
73 Resolution of the President of the CCP Central Committee concerning the measures taken to improve discipline in technical departments and to diminish the number of faulty pieces produced and improve the quality of the final product, of April 27, 1953, in Aša, AZNP, b. 1.
74 *Zápis z aktivu s delegaty Motokovu ze zahraničí zastupci Motokovu se zástupci AZNP konaný dne 11.12.1958* in Aša, AZNP, b. 8a.
75 *Zápis z konference techniků zahraničních zástupců o voze Škoda 1000 MB pořádané v AZNP Mladá Boleslav*, March 14-15, 1966 in Aša, AZNP, b. 8 and 8a.
76 Ibid. and *Zpráva o opatřeních k zavedení prodeje vozu Škoda 1000 MB na vybraných kapitalistických trzích*, April 4, 1964. See also *Problematika trhu a cenová politika v novych podmínkách řízení*, June 6, 1967 in Aša, AZNP, b. 8a. Aša, AZNP, 8a.
77 SÚA, ÚPV-T, 1012, April 16, 1957; December 14, 1957.
78 *Oborová konference pro osobní automobily*.
79 *Export automobilů*, July 5, 1955, the author Dr. Subotka, letter to the main administration of the Ministry of Engineering [*Ministerstvo strojírenství, hlávní správa odbytu*] in Aša, AZNP, b. 55. See also: *Návrh perspektivního plánu oboru osobní automobily od roku 1956 do roku 1975* in SÚA, MAP, b. 200.
80 *Zápis z aktivu s delegaty Motokovu ze zahraničí zastupci Motokovu se zástupci AZNP*, December 11, 1958, in Aša, AZNP, b. 8a.
81 Ibid.

Epilogue

1 Kornai, *The Socialist System: The Political Economy of Communism* (Princeton: Princeton University Press, 1992).
2 SÚA, *ÚPV-T* , b. 48, September 14, 1953.
3 Teichová, *The Halting Pace to Scope and Scale*, 447-61 and Teichova, *The Czechoslovak Economy 1918-1980*, 150.
4 *Problematika trhu a cenová politika v nových podmínkách řízení*, June 1967, Aša, AZNP, b. 8.
5 *Problémy odbytu na vnitřím a zahraničním trhu* and *Výtah z přednášky o marketingu a jeho aplikaci v AZNP Mladá Boleslav*, Aša, AZNP, b. 20.
6 Aša AZNP, b. 8. "Marketing is a total system of interacting business activity designed to plan, price, promote and distribute want-satisfying products and services to present to potential customers." From Giuseppe Volpato, *Evoluzione delle strategie di marketing nell'industria automobilistica internazionale*, "Annali di Storia dell'impresa," 1986, note 6, 126. For studies on marketing in the GDR, see Marcello Anselmo, "Der werktatige Verbraucher: Defining the Socialist Consumer: Market Research in GDR 1960s/ 1970s," in *The Sovietization of Eastern Europe: New Perspectives on the Postwar Period*, eds. Balatz Apor, Peter Apor and Edward Arfon Rees (Washington: New Academia Publishing, 2008), 77-91.

7 Aša, AZNP, b. 65, 5
8 Cohen, *The Soviet Fordson: Between the Politics of Stalin and the Philosophy of Ford, 1920 -32,* and Shpotov, *Ford in Russia from 1909 to World War II*, 531-58 and 525-29.
9 Merkel, *Consumer Culture in the GDR,* and Steiner, *Dissolution of the Dictatorship over Needs?*
10 Aša, AZNP, b. 20.

Foundation for the History of Technology & Aksant Academic Publishers
Technology and European History Series
Ruth Oldenziel and Johan Schot (Eindhoven University of Technology)
Series Editors

The Technology and European History series seeks to present scholarship about the role of technology in European history in the nineteenth and twentieth centuries. The series focuses on how technical communities, nation-states, businesses, social groups, and other actors have contested, projected, performed, and reproduced multiple representations of Europe while constructing and using a range of technologies. The series understands Europe both as an intellectual construct and material practice in relation to spaces inside as well as outside Europe. In particular, the series invites studies focusing on Europe's (former) colonies and on the two new superpowers of the twentieth century: the United States of America and the Soviet Union. Interdisciplinary work is welcomed. The series will offer a platform for scholarly works associated with the Tensions of Europe Network to find their way to a broader audience. For more information on the network and the series see: www.tensionsofeurope.eu

Books in series

1. Judith Schueler, *Materialising identity. The co-construction of the Gotthard Railway and Swiss national identity* (Amsterdam, June 2008)
2. Vincent Lagendijk, *Electrifying Europe. The power of Europe in the construction of electricity networks* (Amsterdam, August 2008)
3. Frank Schipper, *Driving Europe. Building Europe on roads in the twentieth century* (Amsterdam, September 2008)
4. Adri Albert de la Bruhèze and Ruth Oldenziel (editors), *Manufacturing Technology, Manufacturing Consumers. The Making of Dutch Consumer Society* (Amsterdam, January 2009)
5. Irene Anastasiadou, *Constructing Iron Europe. Transnationalism and Railways in the Interbellum* (Amsterdam, March 2012)
6. Milena Veenis, *Material Fantasies. Expectations of the Western Consumer World among East Germans* (Amsterdam, March 2012)
7. Suzanne Lommers, *Europe – On Air, Interwar Projects for Radio Broadcasting* (Amsterdam, April 2012)
8. Valentina Fava, *The Socialists People's Car: Automobiles, Shortages, and Consent on the Czechoslovak Road to Mass Production (1918-64)* (Amsterdam, May 2013)
9. Jiří Janáč, *European Coasts of Bohemia. Negotiating the Danube-Oder-Elbe Canal in a Troubled Twentieth Century* (Amsterdam, November 2012)

Foundation for the History of Technology

The Foundation for the History of Technology (SHT) aims to develop and communicate knowledge that increases our understanding of the critical role of technology in the history of the Western world. Since 1988 the foundation has been supporting scholarly research in the history of technology. This has included large-scale national and international research programs and numerous individual projects, many in collaboration with Eindhoven University of Technology. SHT also coordinates the international research network Tensions of Europe: Technology and the Making of Europe. For more information see: www.histech.nl